Getting the most out of your Abrasive Tools

Deltacraft®

Operating Manuals
and Project Books

GETTING THE MOST OUT OF YOUR DRILL PRESS
GETTING THE MOST OUT OF YOUR CIRCULAR SAW AND JOINTER
GETTING THE MOST OUT OF YOUR ABRASIVE TOOLS
GETTING THE MOST OUT OF YOUR LATHE
GETTING THE MOST OUT OF YOUR SHAPER
GETTING THE MOST OUT OF YOUR BAND SAW AND SCROLL SAW
PRACTICAL FINISHING METHODS
THINGS TO MAKE
ONE EVENING PROJECTS
PROJECTS FOR OUTDOOR LIVING

See the many other books in the Deltacraft Library.

Getting the most out of your
Abrasive Tools

a Deltacraft® Publication

A complete handbook covering all branches of Abrasive tool operation in the home workshop with over two hundred and fifty photographic illustrations and line drawings.

Rockwell **MANUFACTURING COMPANY**
DELTA POWER TOOL DIVISION — PITTSBURGH 8, PENNSYLVANIA

Copyright 1939, 1955
Rockwell Manufacturing Company

Foreword

Getting The Most Out of Your Abrasive Tools is published as a service to power tool users. Different sizes, models and makes of machines vary in their performance, features and ease of operations—and the editors have endeavored to make the information as general as possible.

Abrasive tools are available in a number of sizes, types and designs. Selecting the proper machine for your requirements depends upon the ultimate use you will make of it. For example, you should carefully investigate which type of machine you will need to make sure that it will meet your requirements. The amount you intend to use your abrasive tools is also very important. If your machine will be used for long, sustained periods of time in the home workshop or in metal or wood production work, you should be sure to insist on sealed, lubricated-for-life ball bearings, etc. On the other hand, if your budget is limited and your use will be very intermittent, it may be well to consider lower cost bronze bearing machines.

For your convenience and guidance, the editors have prepared a list of "things to look for" when you buy abrasive tools. This list may be found on page 99 of this manual.

Originally published in 1939, this manual has been reprinted many times and kept up to date by addition of the latest wood and metal working techniques and suggestions. It is the earnest hope of the editors that it will help you "get the most out of your abrasive machines."

DELTA POWER TOOL DIVISION
ROCKWELL MANUFACTURING COMPANY

Contents

CHAPTER ONE—Abrasive Tools   1
The Grinder—The Buffing Head—The Belt Sander—The Disk Sander—Sanding and Grinding Attachments—Mounting Sanding Disks—Fitting Abrasive Sleeves—Sanding Belts—Mounting Grinding Wheels

CHAPTER TWO—Abrasives   6
Natural Abrasives—Artificial Abrasives—Grain Size—Grinding Wheels—Coated Abrasives—Grinding Wheel Selection—Special Types of Abrasives

CHAPTER THREE—Abrasive Tool Dust Collectors   10
Home Made Type—Industrial Type—Standard Motor Driven Unit—Motorless Gravity Unit

CHAPTER FOUR—Operating the Belt Sander   14
Surfacing—End Work—Diagonal Feed—Sanding Inside Curves—Short Work—Use of Sanding Table—Inside Corners—Circle Jigs—Tilting Fence—Pivoted Arm—Beveling Jig—Use of Forms

CHAPTER FIVE—Operating the Disk Sander   22
Freehand Sanding—Pivot Jigs—Rounding Corners—Pointing Dowels—Use of Miter Gage—Grinding Metal—Large Work—Sanding with Pattern—Sanding to Width—Use of Double Disk—Sanding Long Edges—Selection of Abrasive

CHAPTER SIX—General Grinding   27
Safety Suggestions—Odd Jobs—Position of Tool Rest—Use of Guides

CHAPTER SEVEN—How to Sharpen Tools 30
General—Wood Chisels—Honing—Plane Irons—Wood Turning Tools—The Skew Chisel—The Parting Tool—The Gouge—Lathe Tool Bits—Circular Saws—Mortising Chisels—Grinding Jointer Knives—Honing Knives—Jointing Knives—Setting Jointer Knives—Grinding Knives in Head

CHAPTER EIGHT—Grinding Shaper Cutters 46
Rake Angle—Amount of Bevel—Projected Shape—Making a Knife—Use of Shaped Wheels—Sharpening Knives

CHAPTER NINE—Grinding Twist Drills 52
Point Angle—Lip Clearance—Drill Grinding—Web Thinning—Drill for Brass—Special Grinding—Wheels for Drill Grinding

CHAPTER TEN—Drill Grinding Attachment 58
Wheel Dressing—Mounting the Attachment—Adjustment for Lip Angle—Adjustment for Heel Angle—Grinding a Drill—Setting the Jaws—Adjustment for Grinding Position—Wheel Wear—Web Thinning

CHAPTER ELEVEN—Buffing and Polishing 64
Polishing—Polishing Wheels—Setting Up Polishing Wheels—Concerning Glue—How to Polish—Fine Wheeling—Buffing—Strapping Belts—Buffing Compounds

CHAPTER TWELVE—How to Use Sanding Drums 71
Sanding Drums—Sanding on Lathe—Narrow-Face Drums—Sanding on Drill Press—Pattern Sanding

CHAPTER THIRTEEN—How to Use Cut-Off Wheels 76
General Use—Cutting Thin-Wall Tubing—Cutting Solid Stock—True Wheels Essential—Cutting-Off on Grinder—Diamond Blades—Miscellaneous Cut-Off Wheels—Cutting-Off on Lathe

CHAPTER FOURTEEN—Miscellaneous Abrading Operations 81
Tumbling—Spun Finish—Grinding Glass—Engine Finish—Internal Grinding—Other Lathe Operations—Drilling Glass—Grinding Keyways—Surface Grinding—Grinding on Shaper—Sanding on Band Saw—Sanding on Lathe—Sanding on Other Tools

CHAPTER FIFTEEN—Excerpts from Your Delta Craftsheet 92
Spark Test for Metals—Abrasives and Abrasive Terms—Grinding Wheel Selection—How to Sharpen Chisels—(Wood Chisels—Honing)

CHAPTER SIXTEEN—Helpful Hints About Machines and Accessories 99
What to Look for When You Buy Abrasive Tools (Established Manufacturer—Availability of Replacement Parts—Availability of Accessories—Solid Construction—Sturdy Tilting Tables—Adjustable Tool Rests—Safety Features—Correct Speeds) Proper Accessories (Drill Grinding Attachment—Plane Blade Grinding Attachment—Diamond Pointed Wheel Dresser and Tool Holder—Grinding Wheels—Buffing, Wire and Fibre Wheels—Grinding Shields—Abrasive Disks and Disk Adhesive for Disk Sanders—Abrasive Belts for Belt Sanders—Dust Collectors for Abrasive Tools—Miter Gage for Belt and Disk Sanders)

Getting the most out of your Abrasive Tools

Fig. 2

DISC SANDER

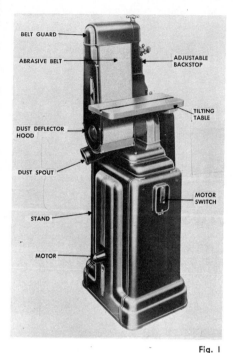

Fig. 1

BELT SANDER

Figs. 1 to 3 show three typical abrasive tools—The Bench Grinder, the Belt Sander, and the Disk Sander. The grinder, a bench model, is direct-motor driven. The belt sander is a medium-size unit using six inch wide abrasive belts. The disk sander is direct-motor drive and uses twelve inch diameter abrasive disks. All units are commonly described according to the abrasive area—a six inch grinder (wheel diameter), a six inch belt sander (belt width), a twelve inch disk sander (disk diameter).

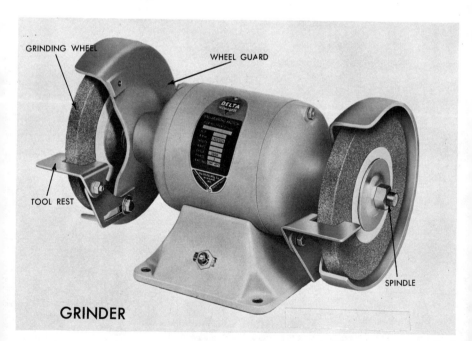

Fig. 3

CHAPTER ONE

Abrasive Tools

The Grinder. The Grinder has a double end horizontal spindle, the ends of the spindle being threaded and fitted with flanges to take the grinding wheels. The spindle is often a continuation of the motor shaft, in which case the unit is direct-driven. Other models employ a conventional belt drive. The size of the grinder is commonly taken from the diameter of the abrasive wheel used in connection

Fig. 5. The 1/3 H.P. bench grinder is ideal for all around grinding, buffing and polishing. It has ample clearance around the wheels, fully adjustable tool rests and eye shields.

Fig. 4. Direct motor driven grinder fitted with safety hoods as well as wheel guards.

with it, that is, a grinder swinging a 7-inch wheel would be called a 7-inch grinder. Units are further described as bench or pedestal, the latter indicating a floor model.

An essential feature of all grinders is the wheel guards. These should enclose the wheel as fully as possible in order to prevent abrasive chips or larger fragments of the wheel from being thrown at the operator. Fig. 6 pictures the results of an experiment conducted to test the guarding qualities of the grinder illustrated in Fig. 4. The grinding wheel was deliberately smashed with a rifle bullet while running at its highest speed (8000 rpm.). The grinder tool rests should be adjustable to allow for wheel wear, and in precision grinders, are also adjustable for tilt. The power required to operate a 6 or 7-inch grinder is approximately ⅓ h. p. Where the unit is direct-driven, the motor must be a 3400 rpm. type in order to give the grind-

Fig. 6 pictures the results of an experiment conducted to test the guarding qualities of the grinder illustrated in Fig. 4.

ing wheel an efficient rim speed. 5500 surface feet per minute is a fair standard for average grinding, although much higher speeds are sometimes used for special work.

The Buffing Head. The buffing head is mechanically similar to the grinder except that guards and rests are not required. A surface speed of about 6500 f. p. m. is suitable for average work.

The Belt Sander. The belt sander features a continuous abrasive belt which works over pulleys at either end of a main sanding table. Adjustments are provided for tensioning and tracking the belt. The size of the unit is commonly designated the same as the width of the sanding belt which it uses. One-half to three-quarter horsepower is required to operate the belt sander. Pulleys should be such as will give a surface speed between 2800 and 3200 feet per minute.

The Disk Sander. The disk sander comprises a circular plate which operates in a vertical position. Cloth or paper backed abrasive disks are cemented or otherwise fastened to the plate. The diameter of the abrasive disk indicates the size of the machine. Two of the most common sizes are the 8½" and 12". Disks of these diameters should run at about 1725 rpm. (standard motor speed). This will give you surface speeds ranging from zero at the center of the disk to about 3835 f.p.m. at the rim of the 8½" size

Fig. 7. The buffing head is similar to a grinder except guards and tool rests are omitted.

and 5500 f.p.m. at the rim of the 12". Materials likely to clog the abrasive should be worked more toward the center of the disk.

Sanding and Grinding Attachments. Accessories for sanding or grinding are used on the drill press, lathe, scroll saw and other machines. The sanding drum, used on the lathe or drill press, is the best known and most used. The surface speed of such drums is best held to a comparatively low figure, say 1200 f. p. m. as com-

Fig. 8

Fig. 9

Figs. 8 & 9 show a light-duty buffing head. Like most similar, belt-drive units, it can be driven from either bottom or back as desired.

pared with an average of about 3000 f. p. m. for long belts. A simple test for efficient speed is indicated by the abrasive drum itself, which will glaze quickly when operated at too high a speed.

Mounting Sanding Disks. In order to present a true, flat abrasive surface to the work, sanding disks are mounted on an accurately-machined metal plate. Glue can be used as the adhesive, in which case the plate and disk must be clamped between boards and allowed to dry overnight. Special types of disk adhesives are available. They are quick acting and will not dry out or cake on your sanding disk. This type of disk adhesive is easy to apply and it holds the disk in place firmly. Fig. 11 shows the disk being cleaned off. Use a screwdriver or blunt instrument for this operation. The disk adhesive is applied as shown in Fig. 12 by holding the adhesive against the disk as it is turning. Adhesive is also applied to the back of the abrasive sheet by laying it on a flat surface and rubbing the adhesive onto the back. The sheet is then placed firmly onto the disk as shown in Fig. 13 and is ready for use, Fig. 14.

Fig. 10. The 8 1/2" disk sander is excellent for small workshop use. It is a multipurpose tool, capable of doing other abrasive operations such as grinding, drum sanding, buffing and polishing.

Fitting Abrasive Sleeves. Abrasive sleeves are mounted on special drums in which alternate layers of rub-

3

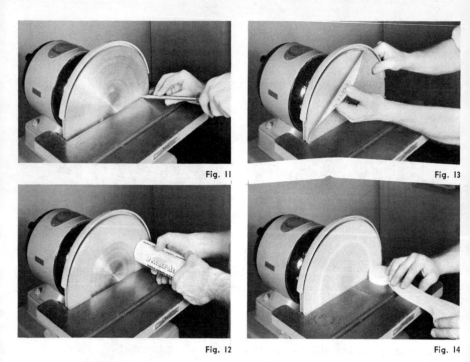

Figs. 11 to 14. *The proper mounting of the abrasive disk is important. Disks must be adhered flatly and smoothly in order to do good work.*

ber and fiber, as shown in Fig. 15, can be expanded by turning the spindle nut, thus securing the sleeve in place.

Sanding Belts. Sanding belts can be purchased readymade for most belt sanders. The worker can also make his own belts by splicing the sandpaper to form a belt of the proper length. Several varieties of belt splices are in common use. The interlocking splice, Fig. 16, is made with an inexpensive cutter. A cloth patch is necessary to retain the two ends in position. This splice is very strong and easy to make, but has the disadvantage of a bump at the joining point caused by the patch. The plain butt splice, Fig. 17, is made without special cutting equipment, the cut ends of the belt being simply patched together at either 45 or 90-degrees. The skived joint, Fig. 18, is the one most commonly used. If one end only of the belt is skived, the joint is a single skive; if both ends are skived, the joint is a double skive. Any suitable angle can be used in making the joint. The grain can be skived from the belt ends by using an abrasive stick about three sizes coarser than the belt. Where a suitable abrasive brick is not available, the skive can be made by dipping the belt end into hot water, as shown in Fig. 19, after which the abrasive and glue can be readily removed from the cloth backing, Fig. 20. The joint is put up with a light coat of medium thick hot glue, and is held in a suitable press until dry, as in Fig. 21.

Mounting Grinding Wheels. Good quality grinding wheels are metal bushed and should be a snug fit on the spindle. Disks of blotting paper should be used on either side of the wheel to serve as shock absorbers. The washer which holds the wheel in place should be of the cup type—never a flat washer. The spindle nut should be turned up securely, and the thread of the spindle must be such that the nut locks against the direction of rotation.

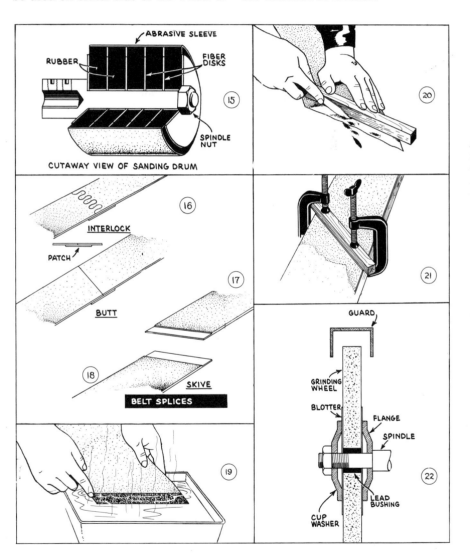

Figs. 15 to 22. Belt splices must be carefully made in order to secure pliability, strength, and a non-bump surface.

CHAPTER TWO

Abrasives

Natural Abrasives. Natural abrasives are found ready made in the earth and include sandstone, emery, flint, garnet, etc. Each has its own particular use. Flint is the least expensive and is the type of abrasive commonly associated with "sandpaper." Garnet is much harder and tougher than flint and is the abrasive most used by the woodworker. Emery is commonly used for sanding metals. For a complete list of abrasives and their characteristics, see page 94.

Artificial Abrasives. Artificial abrasives are a product of the electric furnace. The two main groups are (1) aluminum oxide abrasives, (2) silicon carbide abrasives. Aluminum oxide is made by fusing bauxite, a highly aluminous clay, in an electric arc furnace at about 3,000 degrees F. The crystals are usually brown in color, but some types are made gray and white. They are not as hard as silicon carbide but are much tougher. Silicon carbide is made by fusing sand and coke at a high temperature. The resulting crystals are next in hardness to the diamond, but are brittle as opposed to the toughness of aluminum oxide. The color ranges from black-gray to blue-green. Both aluminum oxide and silicon carbide are sold under various trade names such as Aloxite, Alundum and Lionite (aluminum oxide), and Carborundum, Crystolon and Carsilon (silicon carbide).

Grain Size. The grain size or grit is determined by passing the crushed ore over various wire and silk screens. Fig. 1 shows a 12-grain screen. Grains passed by this screen are called No. 12, twelve grains measuring about 1 inch if laid end to end. Scientific control methods eliminate flat and slivery grains, Fig. 2, unless desired for some specific purpose, retaining only the ideal polyhedral-shaped grain shown in Fig. 3. Sizes range from No. 6 to No. 240. Since it is difficult to make a screen of more than 240 meshes to the inch, finer grains up to No. 600 are graded by an elaborate water flotation system.

Grinding Wheels. Abrasive grains fused with a bond of flux and clay or other substance can be cast into any convenient shape, such as the familiar grinding wheel. Each grain thus becomes a miniature cutting tool, as shown in Fig. 4. As the grains wear down and become dull, they are torn loose from the bond, exposing a new, sharp set of cutting edges. Grinding wheels are made in hundreds of different shapes, a few of the most common styles being the straight, cup and dish wheels shown in Fig. 5.

Coated Abrasives. Abrasive grains glued to sheets of cloth or paper are known as coated abrasives. Disks and sheets, drums, and belts are common examples of coated abrasives. The polyhedral grain shape is always used when coated abrasives are made by ordinary methods, producing a surface similar to that shown in Fig. 6. It can be seen that any method of gravity coating with oblong grains would result in an unsatisfactory surface, many of the grains being almost completely embedded in the glue coating, as shown in Fig. 7. If the oblong grains are placed on end, as in Fig. 8, the result is quite different, the abrasive particles being fully exposed and capable of clean, uniform, high speed sanding. This vertical coating of abrasive grains is done by an electrostatic method, and the greater portion of all coated abrasives used today are made in this manner.

Coated abrasives are divided into many different classes, depending upon the abrasive used, the kind of backing, whether for wet or dry sanding, etc. In any of these, the normal coating is put on in a close, packed formation, hence the general descriptive term "closed coat." The closed coat is fast-cutting and durable, but has the disadvantage of clogging under certain conditions. Where the coating is spaced to show a slight separation between the abrasive grains, the coating is described as "open coat." Open coated abrasives are not as durable as close coated, but they are useful for finishing certain materials where the abrasive dust tends to clog the disk or belt.

In a somewhat similar manner, the grains in a grinding wheel are spaced,

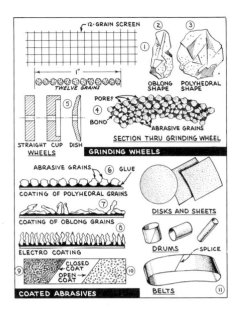

Figs. 1 to 11. Abrasive grains are graded by passing through screens and are then made into various abrasive products.

the word "structure" being generally used to indicate this abrasive spacing. For all average usage, grinding wheels are supplied in a medium structure.

Grinding Wheel Selection. Most grinders are supplied with a general-purpose grinding wheel and this wheel will handle most of the work encountered in the home or small production shop. Where, for any reason, a special wheel is required, the user can arrive at a workable selection by following a few simple rules. Every grinding wheel has five distinguishing features: (1) Abrasive (aluminum oxide, silicon carbide, etc.), (2) Grain (size of abrasive grains), (3) Grade (strength of bond), (4) Structure (grain spacing), (5) Bond (what kind of material used).

CLASSIFICATION OF GRINDING WHEELS

GRAIN SIZES: Number of Abrasive Grains to the Inch.

VERY COARSE	COARSE	MEDIUM	FINE	VERY FINE	FLOUR SIZES
8	12	30	70	150	280
10	14	36	80	180	320
	16	46	90	220	400
	20	60	100	240	500
	24		120		600

GRADE: Strength of Bond.

	VERY SOFT	SOFT	MEDIUM	HARD	VERY HARD
Carborundum	W, V, U	T, S, R, P, O, N	M, L, K, J, I	H, G, F	E, D
Norton	E, F, G	H, I, J, K	L, M, N, O	P, Q, R, S	T, U, W, Z

Fig. 12

The abrasive should be considered first. Only the aluminum oxide and silicon carbide abrasives need be considered. Aluminum oxide is used for grinding all materials of high tensile strength, such as carbon steels, high speed steel, malleable iron, wrought iron, etc. Silicon carbide is used for grinding materials of low tensile strength, such as gray iron, brass and soft bronze, aluminum, copper, etc. Either type of abrasive will, generally speaking, give workable results in either class.

The grain selection is comparatively simple. For soft, malleable materials, a coarse grain gives best results; for hard, brittle materials, the abrasive grains should be fine. The general run of work is done with 60-grit wheels.

The grade of a grinding wheel describes the bond as being hard, medium or soft. Grains are easily loosened from a soft wheel, making it practically self-dressing, while the hard wheel holds together under extreme pressure. Hard wheels are generally used for grinding soft materials, while soft wheels are used for grinding hard materials. For all average work, a medium hard grade will wear well while retaining a sharp edge.

The selection of the proper structure is determined by the nature of the material to be ground. Generally, soft materials, which tend to clog any abrasive wheel, require a wheel with abrasive grains widely spaced. Opposite to this, hard, brittle materials require a wheel with closely-spaced abrasive grains. The wide spacing gives a coarse finish; the close spacing, fine.

Vitrified wheels are the common selection as to bond. Cut-off wheels subject to deflection strains are generally bonded with resin, shellac or rubber.

Special Types of Abrasives. Both aluminum oxide and silicon carbide abrasives are made to special formula other than standard. The pure white aluminum oxide wheel is not as tough as regular aluminum oxide, and hence fractures under mild pressure. This feature prevents overheating, making white aluminum oxide suitable for grinding high speed steel. The principal variation from regular silicon carbide is the green crystal, which features a low order of toughness but is extremely hard and brittle. This feature makes green silicon carbide ideal for cutting and grinding glass and extremely hard alloys. The hardest of all abrasive grains is the diamond, and this type of grinding wheel, made from genuine diamond chips, is extensively used for cutting and grinding glass, and for cutting shapes from blocks of silicon carbide and aluminum oxide.

CHAPTER THREE

Abrasive Tool Dust Collectors

Both the operator and valuable machinery should be fully protected from metal and dust particles produced in harmful quantities by the various abrasive machines in abrading operations. Dust collectors are ideally suited to solve this problem and can be found in many of today's modern shops.

Home Made Type. Many ingenious shop owners have made their own dust collectors for some abrasive machines by adapting old vacuum cleaners obtainable at low cost from community repair shops. While they may be unserviceable for their original domestic purpose without major repairs and replacements, they are still mechanically sound and can be easily adapted to abrasive machines such as belt and disk sanders. Fig 1 pictures a "vacuum cleaner" dust collector which has been installed on a disk sander. The vacuum cleaner and the machine motor should be wired to operate from one switch without necessitating an additional movement to put the "collector" in operation.

The home made type of dust collector should not be used on grinders and other abrasive machines where excessive metal cutting operations are performed. Sparks from the grinding wheels may be drawn into the collector bag by the air stream which acts as a fan causing a hazard should

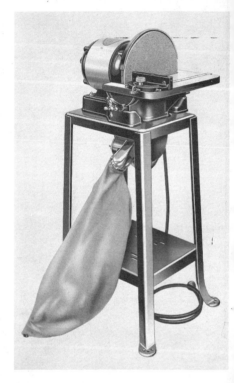

Fig. 1. The home made vacuum cleaner type dust collector.

a combustible material from a previous abrading operation be present.

Industrial Type. Some home workshop owners have gone "all out" by installing a home shop adaptation of the industrial system in which a single blower handles all the waste from the various machines and carries it off to a terminal collector. Fig. 3 pictures

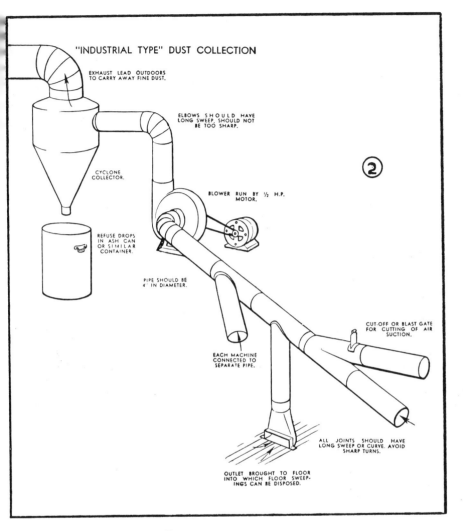

Fig. 2. Industrial type dust collection.

a system of this type.

Standard Motor Driven Unit. The motor driven type is attached directly to each machine. The unit is self-contained, housing a motor, suction fan, filter screen and dust compartment. A ⅓ h.p. motor furnishes the power and is usually located in a separate dust proof compartment. The fan sucks both fine and heavy dust particles through a perfected air filter. The heavy particles fall into a pan at the bottom of the unit and are easily removed. The filter can be removed for cleaning and reused; if properly cared for will last indefinitely.

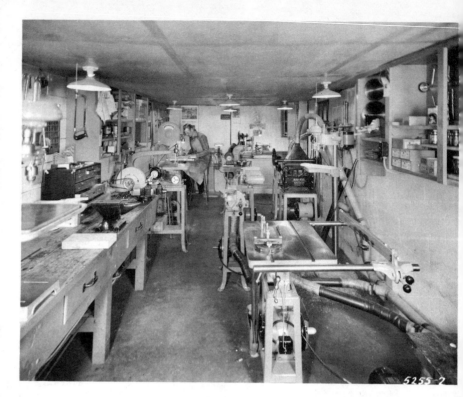

Fig. 3. A home workshop adaptation of the modern industrial type dust collector set up.

Fig. 4   Fig. 5

Figs. 4 and 5. Typical applications of the motor driven dust collector.

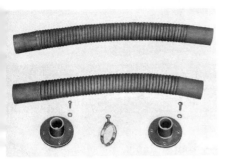

Fig. 6. Special attachment packages can be obtained for easy installation of the motor driven dust collectors.

The inlet openings of the dust collector are located so that the collector can be attached to all types of abrasive tools. Attachment packages consisting of adapters, flexible hoses, hose damps, etc. are available for the various machines. Fig. 6.

Motorless Gravity Unit. The motorless gravity type dust collector like its name implies requires no electrical power to function. It is suitable for abrasive tools such as tool grinders. Fig. 7 pictures the motorless gravity type attached to a grinder. Like the motor-driven type it is a self-contained unit. One advantage to this type of collector is that its operating cost is nil.

Fig. 7

CHAPTER FOUR

Operating the Belt Sander

Surfacing. The sanding table should be in a horizontal position for surfacing. Work can be done freehand, that is, the piece to be surfaced is simply placed on the table. A light but firm pressure should be used to keep the work in the proper position. Excessive pressure against the belt is unnecessary and should be avoided. If

Fig. 1. Surfacing long work.

the work is longer than the table, it is started at one end and gradually advanced in much the same manner as surfacing on the jointer. Where long work is to be surfaced, it is advisable to use the sanding fence as a guide, especially if the board is close to 6 inches wide.

End Work. End sanding is best done with the sanding table in a vertical position, but can be done on the horizontal table by using a guide clamped to the fence, as shown in Fig. 2. The work is pushed down alongside the guide until it contacts the sanding surface. The alignment block permits rapid attachment of the guide at proper right angle position.

Diagonal Feed. The use of a diagonal feed, as shown in Fig. 3, per-

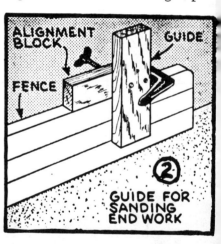

mits the surfacing of work considerably wider than the 6-inch capacity of the belt. The angle of the fence should be kept as small as possible in order to minimize cross grain sanding. A fine belt should be used.

Sanding Inside Curves. Inside curves can be sanded on the end drum, as shown in Fig. 4. The table can be either vertical, horizontal, or at an

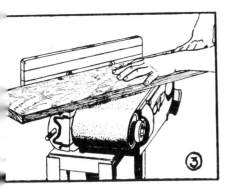

Fig. 3. Use of diagonal feed.

Fig. 5

Fig. 4. Curves are sanded on the outer drum.

Fig. 6

Figs. 5 and 6. Photos show use of backstop.

angle. The fence is used as a guide, being held by one bracket only so it extends beyond the sanding belt.

Short Work. No feed is required on short work up to about 12 inches long, since the full length of such work is in positive contact with a level surface. This permits the use of a backstop to simplify sanding operations. The backstop can be used alone or in connection with the fence, as shown in Figs. 5 & 6. The fence itself can also be used as a stop by swinging it at right angles across the sanding surface.

Use of Sanding Table. Every kind of edge or end work can be done by using the belt sander in a vertical position in connection with the sanding table. With the table level and with the work guided by the miter gage, ends and edges can be sanded true and smooth, either square, mitered, beveled or compound beveled

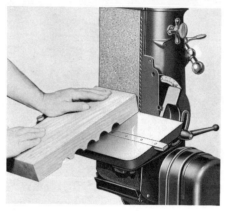

Fig. 7. Using the tilting table.

Fig. 8 shows how inside corners are sanded.

Fig. 9

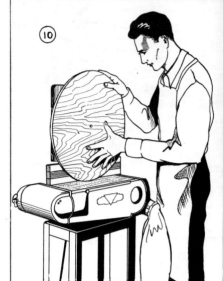

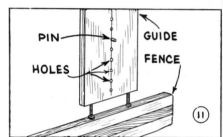

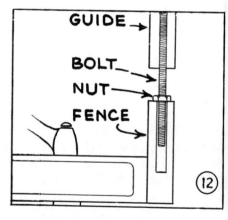

Figs. 9 to 12. Jigs for circles and circular segments can be used to advantage.

is required. A typical operation showing use of the miter gage with tilted table is shown in Fig. 7. Other work is done in the same manner as described in following chapter on use of the disk sander.

Inside Corners. Inside corners can be sanded after tracking the belt so that it runs exactly flush with the edge of the main sanding table. The side guard plate must be removed to permit feeding the work. The work is advanced to the belt alongside the miter gage, as shown in Fig. 8. A cut can be taken on both edges of the corner in one operation, or, each edge can be worked in turn on the flat surface of the belt.

Circle Jigs. All of the various styles of circle jigs using a pivot point can be adapted for use on the belt sander. Fig. 9 shows the use of a pivot arm for segment work. In this form of jig, the work is secured to the forks of

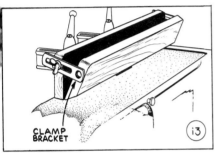

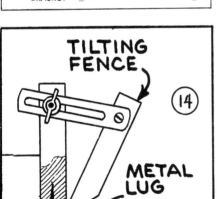

Figs. 13 to 17. A tilting fence and pivot arm are aids to accuracy in production work.

the pivoting arm by means of anchor points. A master form should be used to locate the work at the proper position. Figs. 10, 11 & 12 show a simple jig for sanding circles when the sanding table is horizontal. Holes in the

17

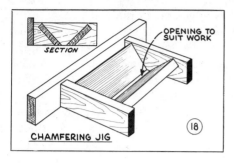

Figs. 18 & 19. The sander affords one of the best methods of working bevels.

fence to take the bolts are for a slide fit so that the nuts can be turned to obtain adjustments for circles of a diameter between the one-inch spacings on the guide board. Normally, circle work is done with the sander in a vertical position, using the sanding table as a support. The same jigs as described for use on the disk sander can be used on the belt sander. Extremely large circles can be worked by mounting the pivot point on any convenient bench or on the table of the drill press.

Tilting Fence. For some types of sanding work, such as beveling, a tilting fence can be used to good advantage. Within certain limits the regular sander fence can be tilted by placing wedges under the bracket arms. More extreme bevels can be sanded by using the simple fence shown in Fig. 13. This consists of a wood fence of the same size as the regular fence. The tilting fence is fastened to the regular fence by means of two metal arms, which are slotted to permit adjustments being made. A thin but rigid strip of metal screw-fastened to the underside of the regular fence, as can be seen in Fig. 14, supports the tilting fence above the sanding belt. One of the metal arms can be fitted with a scale reading in degrees.

Pivoted Arm. A pivoted arm, as shown in Fig. 16, can be used for various jobs where the end of the work is to be pointed or cut off on an angle. The arrangement is simply a fairly heavy piece of wood, which is bolted to the regular sander fence. A washer between the pivoted arm and the fence permits the arm to be tilted or swung to any position. A block clamped to the fence limits the amount of swing while a second block clamped to the arm furnishes a stop for the end of the work. In use, the work is placed on the pivoted arm, one end contacting the stop block. The arm is then tilted down to make the cut, Fig. 17, the operation being completed when the arm comes in contact with the fence stop.

Beveling Jig. One of the cleanest and most accurate methods of beveling, especially on short pieces and end grain, is done with the use of a simple vee-shaped jig. Fig. 18 shows the general construction. The two

Fig. 20. Curved forms of almost any shape can be fitted to the regular sanding table.

pieces forming the vee groove are fitted together at right angles, and are separated at the bottom a suitable distance to make the required cut. Fig. 19 shows the jig in use. The work is simply placed in the vee groove and held there until the sanding belt ceases to cut. If the jig is mounted close to the belt, the width of the bevel will be wider than in the case where the jig is mounted higher. Even wear on the sanding belt is accomplished by moving the fence. Overcutting is impossible, and, providing the jig is parallel with the sanding belt surface, the bevel will be perfectly uniform and straight from one end of the work to the other.

Use of Forms. Sanding in production work can often be done more quickly with the use of forms. These are made from wood to the proper curvature, the form being screw-fastened to the regular sanding table. It is necessary in most cases to make a sanding belt to fit, although a very shallow form can be used with the regular sanding belt. A common shape is the circular form, (see Fig. 22, p. 20). This form is a portion of a true circle, hence, the work can be pushed along it since the curve is the same at all points. Fig. 20 shows work being sanded over a circular form. A suitable fence is made up and clamped to the regular fence, thus providing a side support for the work and insuring square edges. The irregular form, Fig. 23, is not a part of a circle, and work sanded over this type of form must be set down at a certain position, placement being controlled by means of a stop block. Hollow forms, Fig. 24, have curves which do not conform to the belt shape. The belt in this case is run rather loose so that it will be fashioned to the same shape as the form when the work presses against it.

Forms can be built up, the work surface being covered with plywood, as shown in Fig. 21, or the shape may be cut from a solid piece of wood. The preferable abrasive belt for use with forms is the cloth-backed style. A heavy weight backing can be used for most work, the exception being very abrupt curves where a light

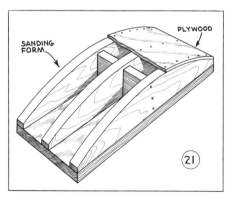

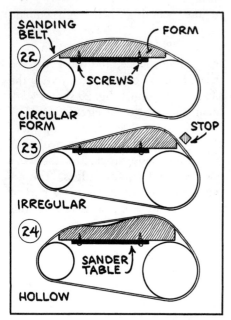

Fig. 26. Using a grooved form to sand a round edge.

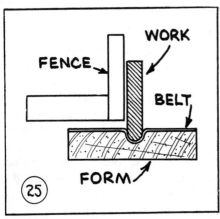

backing gives better results. Forms of the type shown in Fig. 26, always require a light, flexible belt in order to conform to the shape. The manner in which forms of this kind work can be seen in Fig. 25. The fence is aligned with the shaped portion of the form so that the work will be in the proper position when it is pressed against the fence and projected into the belt. Un-

Fig. 27 shows a slashed belt being used to sand abrupt edge curves.

der pressure, the belt takes on the same shape as the form and sands the work to the desired shape. A certain nicety in knowing just when to lift the work from the belt must be acquired by practice. With a lightweight backing surfaced with a fine-grit abrasive, sanding of practically any moulded shape is possible. Where the shape of the form is composed of very abrupt curves, slashed belts should be used. This type of belt, as the name implies, is not a solid surface but is slashed into strips about ⅛-inch wide. Short sections of the belt are left uncut and these uncut portions serve to hold the numerous narrow belts together.

Slashed belts are ideal for sanding odd-shaped edges. This type of work is done without a backing plate, hence either the sander table must be removed or the work done on the back side. The latter is preferable for occasional work since it is much easier to remove the back plate than the table. In use, the belt is run rather slack so that the work, when projected into it, as shown in Fig. 27, will cause the belt to assume the proper shape. Edges finished in this manner will show a very slight curvature, but for all practical purposes the effect is a right-angle cut. Production runs are best done with the use of the sander table, since this provides a rest for the work while insuring proper contact with the belt. It is necessary, of course, to remove the main sander table before the tilt table can be used.

CHAPTER FIVE

Operating the Disk Sander

Freehand Sanding. Sanding on the disk sander is usually done freehand, the work being held flat on the table and projected into the sanding disk. A smooth, light feed should be practiced. Avoid heavy pressure. Best results on curved work can be obtained by going over the work two or three times with light cuts. Sanding is always done on the "down" side of the disk, working on the opposite side would, of course, push the work away from the sanding table. Sanding can be done on the right or left side of the disk depending on the rotation of the motor.

Pivot Jigs. Circular work which is to be sanded should always be worked with the use of a pivot jig. Top and bottom views of a simple jig

Fig.

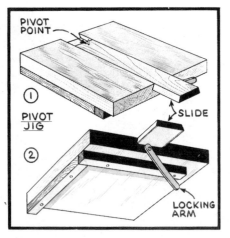

are shown in Figs. 1 and 2. Cleats on the underside provide a positive stop against the front and side of the standard table. The sliding strip can be set at any position, and is locked in place by pushing down on the locking lever, the end of which works like a cam. In use, the work is first band sawed to shape, after which it is mounted on the pivot point. The sliding strip is locked at the required distance from the sanding disk. Pushing the table into the disk sets the cut, and rotation finishes the entire edge to a perfect circular shape. The jig can be clamped to the sander table or simply held with one hand while the other hand rotates the work.

Any other style of pivot jig will work equally well, the simplest set-up being a brad driven into a board which is clamped to the sander table at the required distance from the sanding disk. An overhead pivot point, as shown in Fig. 8, can be made from circular saw hold-down parts. This type of jig is fully adjustable and

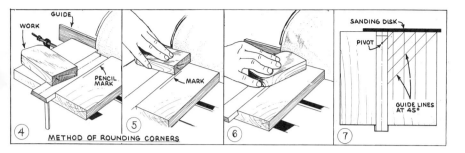

has the advantage of a visible pivot point which can be accurately set in the center of the work.

Rounding Corners. The sanding of corners is allied to circular work in that the edge being worked is part of a true circle. Most work of this nature can be done freehand, sweeping the corner of the work across the face of the sanding disk two or three times until the desired round is obtained. More accurate results are possible if the pivot jig is used in the manner shown in Figs. 4 to 7. The sliding strip is first locked in place at the required distance from the face of the sanding disk. A pencil mark is then drawn on the table of the jig, this mark being the same distance from the pivot point as the pivot point is from the sanding disk, as shown in Fig. 4. The work is placed against a guide fastened to the rear edge of the jig as shown in Fig. 5, and is brought down on the pivot point in alignment with the pencil mark. Rotating the work rounds the corner, see Fig. 6. Fig. 7 shows how the jig table can be marked with pencil lines as a guide to placing work of any radius.

Fig. 8. A *pivot jig* is almost a necessity in sanding circular pieces if accurate work is to be done.

Fig. 9. Pointing dowels on the disk sander.

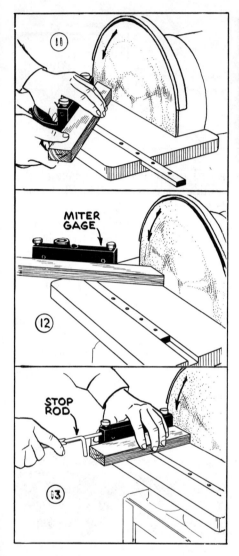

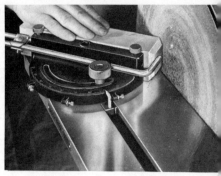

Fig. 14. Square posts are easily beveled by using the gage and stop rod as shown.

Figs. 11 to 13. How to use a miter gage

Pointing Dowels. A hole of the same diameter as the dowel stock is drilled through a scrap piece of wood which is clamped to the sanding table at the required angle, as shown in Fig. 9. The work is pushed through the hole until it contacts the sanding disk, after which it is rotated to finish the point.

Use of Miter Gage. A circular saw miter gage can be used to advantage in sanding square or mitered ends. Where miters are being sanded, the preferable position is as shown in Fig. 11, which permits better handling than the reverse position shown in Fig. 12. Square ends are sanded by projecting the work along the miter gage until it contacts the disk. Sanding to exact length can be done by pre-setting the stop rod at the required distance. The rod is free to slide in the hole in the end of the gage, as shown in Fig. 13, the exact length being set when the rod comes to a stop at the bottom of the hole. The beveling of square posts is easily done by using the miter gage with stop rod in the manner shown in Fig. 14.

Grinding Metal. Finishing metals and plastics on the disk sander is practically the same as similar operations on wood with the exception that

Fig. 17. *Using a spacer pin to sand curved work to exact width.*

Fig. 15. *Clean, accurate work in production runs can be done by sanding with the use of a pattern.*

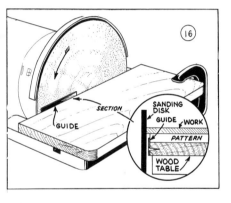

an aluminum oxide abrasive disk should be used instead of garnet.

Large Work. Where the work being sanded is so large that it cannot be easily held on the sander table, an auxiliary wood table of suitable size should be made, this being clamped or otherwise fastened to the standard sanding table.

Sanding with Pattern. In production work, sanding with the use of a pattern can be used to advantage and insures perfect work. A wood table, to one side of which is screw-fastened a thin but rigid strip of metal, is clamped in place over the regular sanding table, as shown in Fig. 16. The guiding edge of the metal strip should be about ⅛ inch from the surface of the sanding disk, and the pattern should be made ⅛ inch undersize to correspond. Anchor points permit fastening the pattern to the work, after which the work is band sawed about 3⁄16 inch outside the edge of the pattern. The work is then sanded smooth, the pattern being held in contact with the metal guide, as shown in Fig. 15, as the work is projected into the sanding disk.

Sanding to Width. Curved surfaces can be sanded to uniform width by first band sawing and smoothing one side, and then using a spacer pin, as shown in Fig. 17, to set the finish cut on the opposite side. While the photo shows the pivot jig used for this purpose, it can be seen that any scrap piece of wood with one corner rounded can be clamped to the sanding table to serve as a guide.

Use of Double Disk. When working small wooden or plastic parts requiring two grades of abrasive for finishing, good use can be made of a double sanding disk. This is made by cutting out the center of the coarser disk, cementing a smaller disk of finer abrasive in the opening thus provided. The work is first sanded on the outer portion of the disk and then without stopping the machine, is finished by means of the finer abrasive disk.

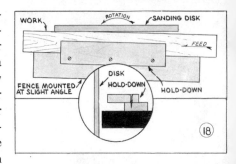

Sanding Long Edges. While the disk sander is not particularly suited for sanding long straight edges, good work can be done by using the set-up shown in Figs. 18 & 19. A wood fence to which is fastened a hold-down block is clamped to the sander table at the required distance from the sanding disk. The wood fence should be mounted at a slight angle to make sure that the work, which is fed from the "up" side of the disk, will not come in contact with the sanding surface until it reaches the "down" side. The angle should be very slight and is purposely exaggerated in the illustrations to show the method of working. A smooth feed is essential. Any length of work can be handled in this manner, or, short pieces can be run through one after another.

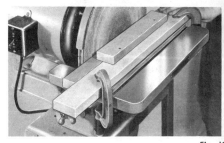

Figs. 18 & 19. Method used in sanding long, straight edges.

Selection of Abrasive. The abrasive used on the disk sander will depend upon the work. As on all other abrasive machines, garnet is used for wood while aluminum oxide and silicon carbide disks are used for metal. Since the disk sander is commonly employed for edge work, the abrasive generally can be somewhat coarser than for surfacing. A ½ or 1/0 disk cuts rapidly to a fairly smooth surface. Fine cabinet work, however, requires final sanding with a 2/0 or 3/0 disk so abrasive scratches will not show.

CHAPTER SIX

General Grinding

Safety Suggestions. The grinder is a safe tool to operate providing a few simple rules are followed. Always use the guards. If guards are not provided, wear suitable goggles as a protection against flying fragments of abrasive. Keep the wheels round by dressing whenever required. Do not force work against a cold wheel, but exercise light pressure until the wheel becomes warm. Always use a tool rest when the work permits. Present the work to the wheel either straight in or at a "drag"

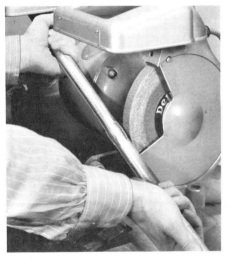

Fig. 1. The grinder plays an important part in hundreds of odd jobs around the shop. Smoothing welded joints and cutting sheet metal are typical examples.

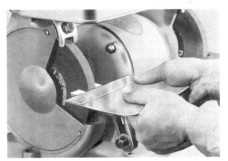

Fig. 3

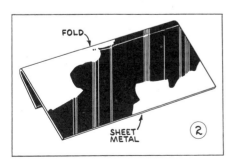

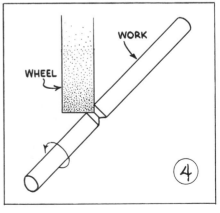

27

angle, reserving the "gouge" angle for sharpening and other operations demanding a minimum burr.

Odd Jobs. An almost endless number of odd jobs are done on the grinding wheel. Smoothing a welded joint, as in Fig. 1, is typical of this class of work. Cutting thin metal by first folding it, as in Fig. 2, and then grinding through the fold, as in Fig. 3, is another example. Fig. 4 shows how the edge of the wheel is used to nick rod stock prepatory to chiseling or sawing. Replacing broken shovel handles is still another job in which the grinder plays an important role. Figs. 5 and 6 show its use. Most offhand grinding is done on the face of

Fig. 5. The grinder with the tool rest removed is used to remove old rivets from a broken shovel.

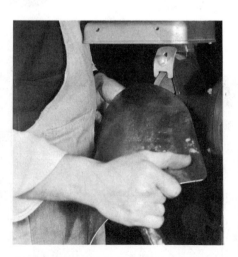

Fig. 6. The cutting edge of a damaged shovel is being sharpened using a grinder.

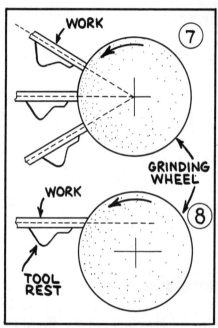

Fig. 9

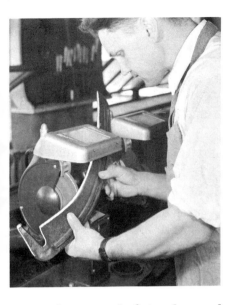

Fig. 10 shows a scythe being sharpened on the pedestal grinder.

Fig. 11. The grinder is ideal for sharpening hatchets. The adjustable tool rest is set at the correct angle before passing the blade over the face of the grinding wheel.

the wheel. When grinding is done on the flat sides of a straight wheel, use care to wear the wheel smooth since a rough or grooved side can be very dangerous on certain jobs.

Position of Tool Rest. A level tool rest set a little below the center of the wheel, as shown in Fig. 7, is in the most practical and safest position for general work. Work ground in this position, or any other position when the work points to the center of the wheel, will be finished with a square edge. It can be seen, Fig. 8, that work presented in any position other than pointing to the wheel center, will be ground more or less on a bevel. Freehand grinding without the use of a rest should always be done on the lower quarter of the wheel.

Use of Guides. Guides clamped to the regular rest insure accuracy and should be used on all precision work. Fig. 9 is an example. The exact bevel and depth of cut is controlled by means of the simple fence against which the work is placed.

CHAPTER SEVEN

How to Sharpen Tools

General. Two operations are necessary in sharpening most tools: (1) the edge is ground to the proper shape on the grinder, (2) the edge is honed to perfect sharpness on a suitable oilstone. The grinding wheel used should be an aluminum oxide wheel, about 60-grit, and of medium hardness. Keep the wheel properly dressed. A revolving disk type dresser (see page 48) can be used satisfactorily. In grinding, keep the tool cool by constantly dipping in water; temper is being drawn when blue spots appear on the edge of the tool. High speed steel is best ground entirely dry, using a very light feed and stopping between cuts to allow the tool to air cool. The use of a white aluminum oxide wheel will permit a heavier feed without overheating.

Wood Chisels. Wood chisels should be hollow ground. Project the chisel straight into the wheel to re-

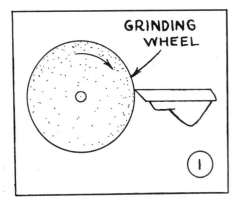

Fig. 1. Remove nicks by pushing chisel straight into wheel.

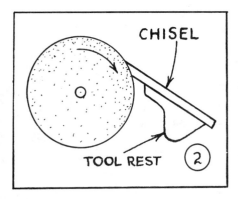

Fig. 2. Tilt tool rest to grind bevel to the required angle.

Fig. 3. Tool grinding is done on an aluminum oxide wheel, with the edge of the tool against the direction of rotation.

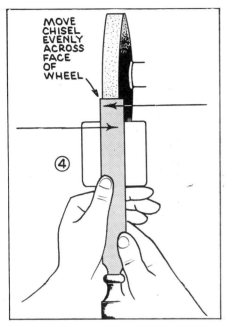
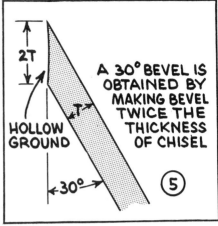
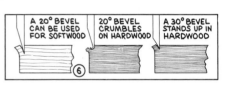

move nicks, as shown in Fig. 1; then, adjust the tool rest to the required position to grind the bevel, Fig. 2, working the chisel squarely across the face of the wheel, as shown in Fig. 4. Worked on the face of the wheel, the bevel will have a slight hollow, making it easy to hone to a perfect edge several times before regrinding again becomes necessary. The bevel should be about 30 degrees, this being obtained by making the bevel twice the thickness of the chisel, as shown in Fig. 5. A 20 degree bevel can be used for softwood, but this thin wedge will crumble on hardwood, as pictured in Fig. 6.

Honing. Either an aluminum oxide or a silicon carbide oilstone will give good results in honing or whetting the chisel edge after grinding. The sharpening stone should always be oiled, the purpose of this being to float the particles of metal so that they will not become embedded in the stone. Use a thin oil or kerosene. Wipe the stone after using. Honing is necessary because grinding forms a burr at the chisel edge, as shown in Fig. 7. To remove the burr, place the chisel diagonally across the stone, as shown in Fig. 8, and stroke backward and forward, bearing down with both hands. The heel of the chisel should be a slight distance above the surface of the stone, as shown in Fig. 9. Next, turn the chisel over and stroke the back on the stone, making certain to keep the tool perfectly level, as shown in Fig. 10. Alternate the honing on bevel and back until the burr is completely removed.

It will now be noted that honing puts a secondary bevel on the chisel,

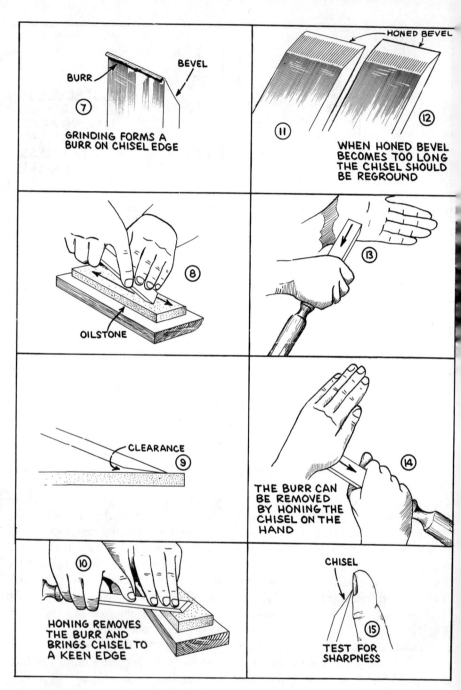

Figs. 7 to 15. Drawing shows honing methods.

as shown in Fig. 11. This is the correct technique for chisels, plane irons, knives, etc. This method gives a clean edge with a minimum amount of labor. When the honed bevel becomes too long through repeated whettings, Fig. 12, chisel should be reground. Figs. 13 and 14 picture the common method of hand honing. Fig. 15 shows standard test for sharpness —the chisel should "bite" on the thumb nail.

Plane Irons. Plane irons are sharpened the same as wood chisels.

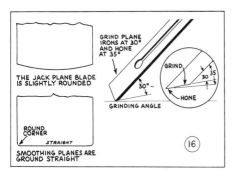

Fig. 16. Jack plane irons should be slightly rounded; all others are ground straight with a slight round at the corners.

A bevel of 30 degrees and a honing angle of 35 degrees is satisfactory in most cases. The corners of the plane iron should be slightly rounded. Mechanical devices to hold the plane iron or chisel at the proper angle when grinding are advantageous. Fig. 18 pictures a plane blade grinding attachment. A carriage table is clamped to the adjustable arm on the grinder. The table tilts so that any desired angle may be obtained. The adjustable quadrant is located on the left side which permits accurate grinding of angle knives as well as assuming an accurate right angle position for straight knives. The attachment has a knurled adjusting nut which allows you to control the amount of "cut" when grinding. It will take knives up to 3-$9/16$ inches wide. A diamond pointed wheel dresser may also be clamped in the holder for wheel dressing.

Wood Turning Tools. Some wood turning tools are not hollow ground. Instead, the bevels are perfectly flat and should be kept flat during honing. Any secondary bevel on a

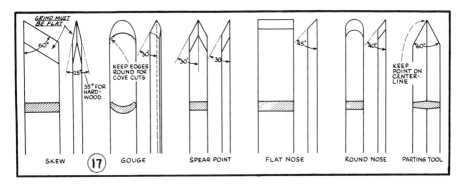

Fig. 17. Grinding angles for wood turning tools.

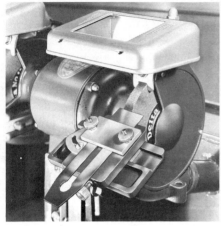

Fig. 18

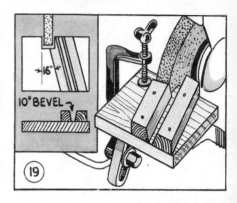

Fig. 20

Fig. 21

skew, for example, will prevent it from being used satisfactorily in turning, where the heel of the bevel must act as a fulcrum.

The Skew Chisel. The bevel of the skew chisel is double and has an included angle of 25 to 35 degrees. Grinding can be done on the disk sander, or on the side of a straight or recessed grinding wheel. The chisel can be held freehand, but better results will be obtained with a simple guide block, as shown in Fig. 19. Holding the chisel first against one of the beveled guide blocks and then the other, as shown in Figs. 20 & 21, will bring each of the bevels to the required angle. In honing, maintain the same bevel.

The Parting Tool. Make a suitable guide block to present the parting tool to the side of the grinding wheel. A grinding wheel mounted on the circular saw offers a convenient method of working since the bevel angle can be set on the miter gage as shown in Fig. 22. The parting tool can also be hollow ground if desired.

Figs. 19 to 21. Bevels on wood turning tools should be ground flat. Simple jigs simplify the work.

The Gouge. Simplest of all methods of sharpening the gouge is to use a cup wheel on the lathe, rotating the chisel inside the wheel, as shown in Fig. 23. The curved surface of the

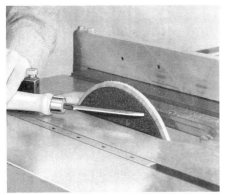

Fig. 22

Fig. 23

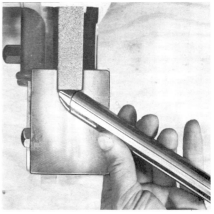

Fig. 24

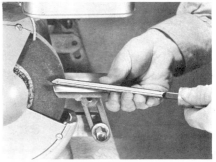

Fig. 25

cup wheel lessens the amount of rolling necessary and makes grinding quite simple. Lacking the cup wheel, the same general method can be employed by turning a wedge-shaped recess in a block of hardwood. Fed with 60-grit abrasive grains combined with grease, the wood block works just the same as cup grinding wheel. The gouge can also be ground by rolling the bevel on the face of the wheel, as in Fig. 24 or on the side of the wheel, Fig. 25. In all cases the roll must be just a little less than that of a full half-circle.

Special sharpening stones are required for honing the gouge. The best type is hollow on one side and round on the other side—made especially for honing gouges. Fig. 26 shows this stone in use. The round edge of the stone can be conveniently used to cut the burr on the inside flat edge of the gouge, or a slip stone can be used for this purpose, as shown in Fig. 27.

Lathe Tool Bits. Lathe tool bits are sharpened offhand, being held in the hand to present the tool at the proper angle to the side of the grinding wheel. A light touch on the grinding wheel is all that is usually required to bring the bit to a keen edge. Maintain the original bevels,

Figs. 23 to 25. The gouge is sharpened by rolling on either a cup or straight wheel.

or, if working blank stock, follow the angles for ¼ inch bits given in the drawing below. Handling during grinding is simplified if the bit is held in the tool holder, as shown in Fig. 28. Small abrasive sticks can be used to advantage for touch-up sharpening, especially when this is done during the course of a lathe job.

Circular Saws. Saw blades can be gummed free hand by simply holding the saw to the wheel. A pencil mark should be made around the rim of the saw, as shown in Fig. 35, to indicate the depth to which the gullets are to be ground. Grinding is done on a narrow, round-edge wheel.

An automatic set-up for grinding and gumming can be made on the

Fig. 26

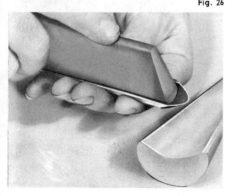

Fig. 27

Fig. 28. Lathe tool bits are ground freehand, the bit being mounted in the tool holder as shown.

lathe with the use of the slide rest, as shown in Fig. 33. The grinding wheel should be dressed to the required gullet shape, and the slide rest adjusted so that the wheel will be in alignment to cut a quarter-pitch tooth, as shown in illustration, Fig. 35. Grinding is done by feeding the saw into the wheel by means of the slide rest feed. A nail in the work table provides a stop so that each tooth is accurately aligned for grinding. A clamp fitted across the slide rest base provides a stop for depth. Each tooth is ground in turn. The saw should be free from gum which might cause an inaccurate setting against the guide pin. After the gullets and tooth faces are ground, the position of the slide rest can be changed to grind the backs of the teeth. Throughout the whole operation of sharpening the circular saw by grinding, caution must be used to prevent burning.

Mortising Chisels. The internal surface of the mortising chisel should be ground to an included angle of 78 degrees. The simplest method of working is to use a small grinding wheel with end shank, dressing the

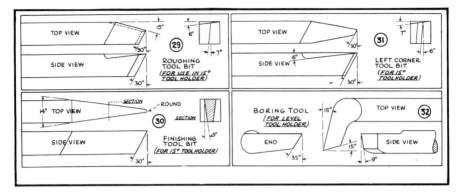

Figs. 29 to 32. Grinding angles for lathe tool bits. (Best results will be obtained by using exact shapes shown)

Fig. 33

Fig. 34

wheel to the required angle. Lacking the wheel, a wood form can be turned and then coated with glue and rolled in abrasive, or, the form can be covered with 120-grit wet-or-dry abrasive paper. The method of working is shown in Fig. 36, the chisel being centered on the lathe tailstock center and projected into the revolving wheel or abrasive-covered wood form. A square abrasive stick or file can be used to clean out the corners.

Grinding Jointer Knives. Jointer knives are ground on an angle of 36 degrees, as shown in Fig. 38. When mounted in the cutterhead, the rear edge of the bevel should be about $1/16$ inch from the surface of the cutterhead, as in Fig. 39. Since the knives are quite narrow, it is necessary for grinding purposes to make a holding block, this being made by running in a saw cut with the block in one of the positions shown in Fig. 40, depending on the method which will be used in grinding. The saw kerf usually will be a snug fit for the knife, but if any looseness is apparent, the knife can be held secure by means of screws.

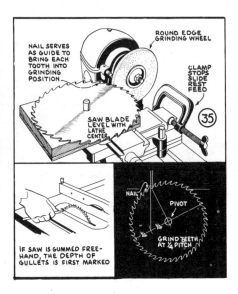

Fig. 35. Simple automatic method of gumming and grinding circular saw blades.

Figs. 36 to 37. Grinding a mortising chisel on the lathe.

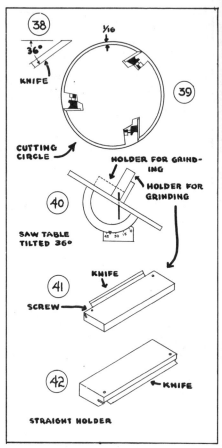

Fig. 43 shows one of the knives being sharpened on the grinder. The tool rest is adjusted to the required angle, and a guide block is clamped in position to insure a straight cut across the knife. Each knife is worked in turn, making a single, very light cut. A strip of paper is then pasted to the holding block, the purpose of this being to set the next cut without changing the original position of the guide block. Two or three very light cuts will usually bring all of the knives to a perfect edge. It cannot be stated too strongly that abrasive cuts on high speed steel knives must be light; heavy cuts will invariably burn the knife and render it useless.

The method of grinding jointer knives on the drill press with a cup wheel is shown in Fig. 44, while Fig. 45 shows the operation as performed

Fig. 43

Fig. 44

Fig. 45

Figs. 43 to 45. A number of different methods can be used in grinding jointer knives. In each case, grinding is done dry. Very light cuts must be taken to avoid burning. Cuts are made in rotation on all knives until a final cut brings each of the knives to a perfect edge.

on the circular saw. The latter method can also be used in connection with the belt or disk sander. Whatever method is used, best results will be obtained if the grinding is done with successive light cuts, taking each knife in turn until all edges come up sharp.

Honing Knives. Grinding is not always necessary to sharpen the jointer since careful honing at regular intervals will maintain a sharp head for some time. To hone the knives, partly cover a fine carborundum stone with paper so it will not mark the table, and place it on the front table, as shown in Fig. 46. Turn the cutterhead until the stone rests flat on the bevel, and fix the head in this position by clamping the belt to the stand. Whet the knife by stroking the stone lengthwise with the blade, treating each knife with the same number of strokes.

Many craftsmen have constructed an attachment which is more elaborate for honing jointer knives. (Fig. 47.) The jig is made of hardwood, either birch or maple, and uses a standard 3″ diameter straight edge honing stone. Figs. 48, 49 and 50 detail the construction of the jig. For the workshop enthusiast who takes pride in his machines, this attachment will help keep his jointer in tip-top shape.

The jig is used as follows:

Raise the front table $1/16$ inch above the rear table. Adjust the stone holder on rear table as shown in Fig. 51, make sure to press stone firmly on table before tightening clamp screw of jig. After the jig has been adjusted, lower the rear table

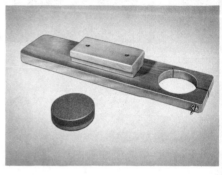

Fig. 47. The finished jig used a standard 3″ diameter straight edge honing stone.

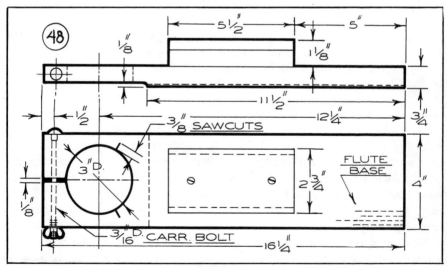

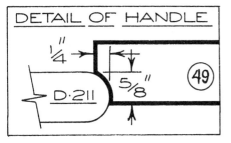

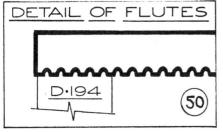

until level with the back edge of the knife, check this with a straight edge.

Place scale on front table and adjust height until end of scale seats itself against knife as shown in Fig. 52 and the line drawing, Fig. 53. Then clamp scale to table with a "C" clamp. (Figs. 52 and 53.)

Two ways of holding the knife firmly against the scale is with a "C" clamp fastened to the V-Belt as shown in Fig. 52, or with a rubber band fastened on to an Allen wrench in the pulley and the other end held on the extension casting of the front table. (Fig. 54.)

Before using the stone saturate it with kerosene. Hone each knife with a circular motion.

Without removing scale, rotate the head backward until scale drops over the next knife to be honed. Do this same thing on the third knife and hone the knives in the same manner.

Remove burr or wire edge with a piece of hardwood as shown in Fig. 55.

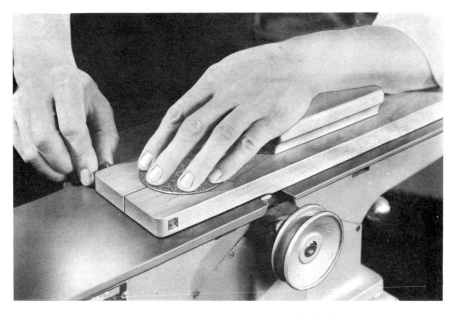

Fig. 51. Here the stone is being adjusted in the jig and locked in place.

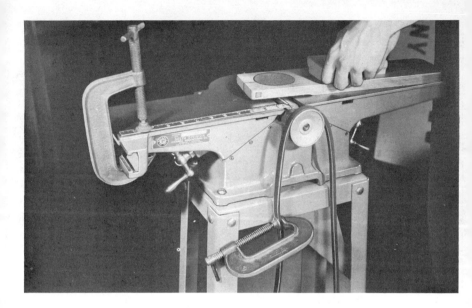

Fig. 52. The jig is in a honing position and is used in a circular motion. Note clamp on scale and V-belt.

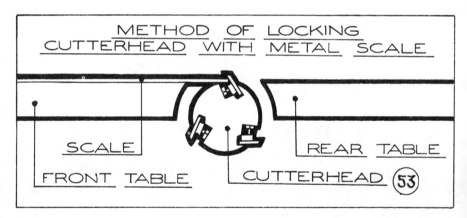

To distribute even wear on the stone, rotate it ⅓ turn for each knife honed.

Jointing Knives. Knives can be sharpened and brought to a true cutting circle by jointing their edges while the head is revolving. In this operation, the stone is placed on the rear table, as shown in Fig. 56, and the table lowered until the stone barely touches the knives. After two or three jointings of this nature, it will be necessary to recondition the knives by grinding in order to maintain back clearance.

Setting Jointer Knives. After grinding, knives must be carefully mounted in the head. One of the best

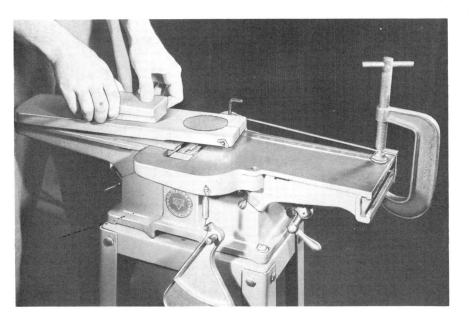

Fig. 54. Another view of the fixture in use. Note rubber band on Allen wrench and around casting of front table.

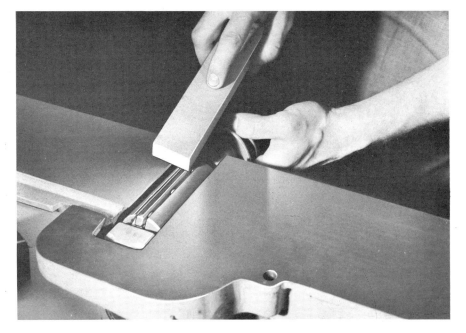

Fig. 55. For a keen cutting knife always remove the burr with a hardwood block.

Fig. 58

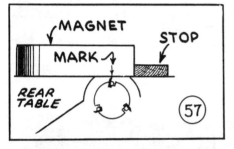

methods of doing this is with a magnet, as shown in Figs. 57 & 58. An index mark should be scribed on the magnet and a stop block should be clamped to the front table at such a position as to bring the index mark in line with the cutting edge of the knife when it is at its highest point. The knife is placed in its slot and is pulled up to the required level by the magnet, after which the setscrews are tightened. Once the initial set-up has been made, this method of ad-

Fig. 59. Grinding jig in use.

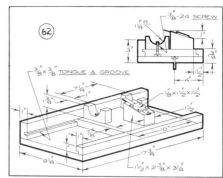

Fig. 60. The jig with jointer head removed.

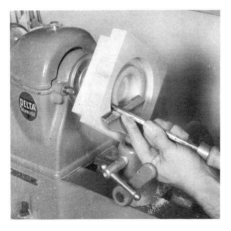

Fig. 61. Pillow boxes should be turned to exact diameter.

justing knives will be found both accurate and convenient, and faster than where a plain straight edge is used.

Grinding Knives in Head. Knives can be ground without removal from the head by the method shown in Figs. 59 to 62.

The complete head is mounted on a sliding jig. A spring clip held in a block serves to bring each of the knives into the required position for grinding. The stationary part of the jig is bolted to the drill press table, grinding being done by sliding the top portion of the jig in a back and forth motion. The entire jig is made of hardwood (birch or maple). For exact seating of the bearing housings, the pillow blocks can be turned perfectly round in the lathe, (Fig. 61). The drawing Fig. 62 details the construction of the jig.

Another method of grinding the complete head is shown in Fig. 63. This makes use of a flexible shaft fitted with a small grinding wheel. The head is fixed at the proper position to maintain the bevel by clamping the belt to the side of the machine stand.

CHAPTER EIGHT

Grinding Shaper Cutters

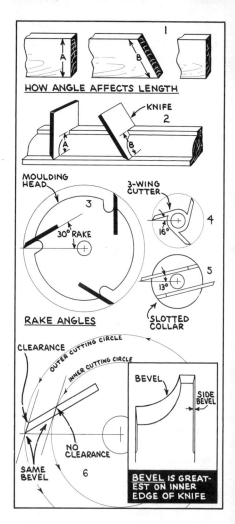

Figs. 1 to 6. Factors governing the knife shape.

Rake Angle. The rake angle of any cutter determines its shape and other characteristics. As shown in Fig. 1, a slanting line across a piece of wood is necessarily longer than a straight one. Applied to shaper cutters, it can be seen, Fig. 2, that the length of the cutter working on an angle, B, must be greater than if the cutter worked straight across the work, as at A. This rake angle is present in all shaper cutters and is greatest when knives are mounted in a moulding head, as shown in Fig. 3, where the angle is approximately 30 degrees. It is obvious that the greater the rake angle, the greater the difference between the shape of the knife and the moulding it cuts.

Amount of Bevel. Knives are beveled at an angle between 30 and 45-degrees. It can be seen, Fig. 6, that a bevel which will provide clearance at the outer cutting circle may not be enough to give clearance at the inner cutting circle. Examination of a factory-sharpened cutter will show that the bevel is greatest at the inner edges of the knife, thus maintaining the same amount of clearance. Portions of the knife parallel with the line of travel, such as the sides, demand only a minimum amount of bevel to provide clearance.

Projected Shape. The required shape of any cutter to produce a certain shape can be obtained by drawing the moulding full size on a piece of paper, as shown in Fig. 7. Along the edge of the moulding erect a vertical line, A. Below the moulding, draw a horizontal line, B, and where A intersects B draw a line, C, at the same angle as the rake angle of the cutter. Drop lines from the moulding shape to the line C, and, using O as a center, carry these lines around to line B on the opposite side and then project them upward. Lines D located at all points where the vertical lines cut the moulding are carried across the center line and establish a series of marks, which, when joined, show the shape of the cutter required to cut the moulding. The difference

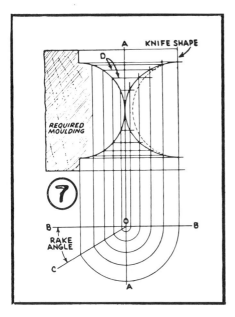

Fig. 7. Diagram shows knife projection drawing.

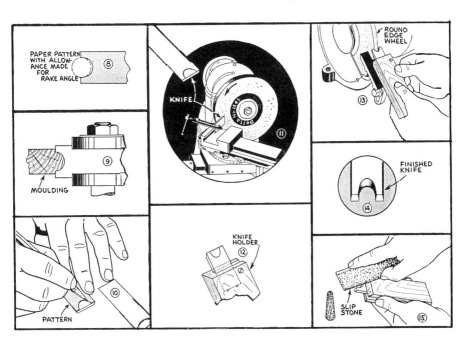

Figs. 8 to 15. Drawings show knife making.

47

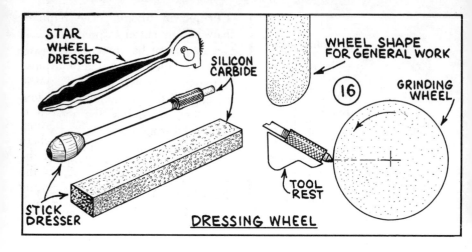

Fig. 16. A star wheel dresser is fast-cutting while stick and diamond dressers provide for precision work.

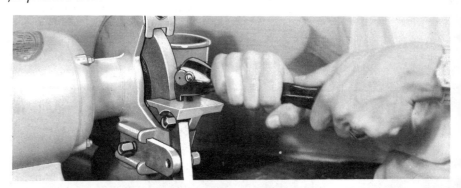

Fig. 17

Fig. 18. The diamond wheel dresser is being used with the wheel dressing tool holder.

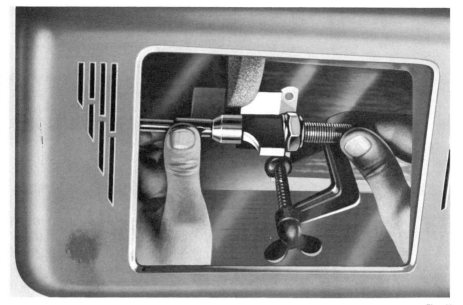

Fig. 19

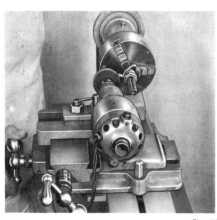

Fig. 20

amounts to about $\frac{1}{16}$ inch in depth where moulding head cutters are being plotted and about half of this for three-wing cutters and slotted collars. For average work, the projected shape can be judged with fair accuracy without drawing. The basic rules to remember are (1) knives for cutting beads must be ground deeper, and (2) knives for cutting coves must be ground fuller.

Making a Knife. Figs. 8 to 15 on page 47 show the various steps in making a pair of knives for use with slotted collars. The required shape is a full half-circle, as shown in Fig. 9. A paper or metal pattern is made, as shown in Fig. 8, and, following the basic rule, this is cut slightly deeper than the shape of a true circle. Fig. 10 shows the pattern shape being transferred to the knife blanks. The outer straight bevel is then ground, a suitable method being as shown in Fig. 11 which uses the lathe slide rest to set the required angle. The curved portion of the knife is then ground on a round edge wheel, as shown in Fig. 13, the tool rest being adjusted to provide the proper bevel. After grinding both knives, the shape is compared and checked, readjustments made as re-

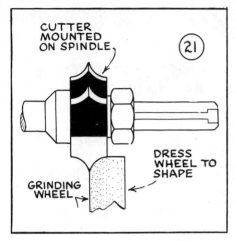

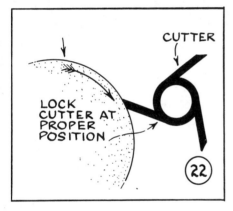

mond dresser. The wheel type should be pushed straight into the wheel, while the stick or diamond work best at a drag angle, as can be seen in Fig. 16. With a wheel properly shaped, it is a simple matter to grind any cutter to the same contour. If precision dressing is required, the diamond pointed wheel dresser, Fig. 18, should be used. It is excellent for all types of general grinding wheels. A wheel dresser tool holder is also available for use with this tool and provides the secure bearing needed in dressing operations. Fig. 19 shows a three-wing cutter being ground. Stops and guides insures all wings being ground exactly the same. The use of a shaped wheel in a tool post grinder used on the lathe is shown in Figs. 20, 21 & 22. The cutter is turned to the required position for the bevel, as in Fig. 22, and is then locked in this position by means of the index pin, after which the cut can be made.

Sharpening Knives. Factory-ground shaper knives with involute bevels should be sharpened by honing the flat side of the cutting edge, quired, after which the bevel is lightly honed, as in Fig. 15, to remove any burr left by the grinding. On certain shapes, good use can be made of a cut-off wheel to remove excess knife stock, thereby eliminating tedious grinding.

Use of Shaped Wheels. Wheels can be fashioned to any required shape by using a suitable dresser. The revolving wheel type is the fastest cutting, but does not permit the precision which is possible with the silicon carbide stick type or the dia-

Fig. 23

as shown in Fig. 23. The involute bevel will retain the same shape regardless of metal removed from the back side. Knives ground in the home-shop with a straight bevel can be resharpened in the same way, or, the bevel itself can be honed. Where the knife has an involute or curved bevel, however, no grinding or honing should be done on the bevel.

CHAPTER NINE

Grinding Twist Drills

Point Angle. The two most important features of drill grinding are (1) the point angle, and (2) the lip clearance. The point angle has been established at 59 degrees for general work, and this angle should be maintained. It is easily checked with a drill-point gage. Gages in a variety of styles can be purchased at a nominal cost, or, the worker can make his own from sheet metal. The markings on the edge, which can be seen in Figs. 1, 2 and 3 need not be exact since they are used only to check the length of one lip against the other. In use, the drill body is held against the edge of the gage, and in such a position that the angular edge is over the cutting lip of the drill. The gage will then show whether or not the point, or, rather, one edge of the point, is "on the 59." Fig. 1 shows a drill with the correct point angle. Fig. 2 shows drill with point angle which is too great; Fig. 3 shows a drill with a point angle too small. Besides being ground to the correct angle, both lips must be exactly the same length. What happens when the lip angles are different or of unequal length is shown in Fig. 4 and Fig. 5. It can be seen that the resulting hole will be out of round and larger than the drill.

Lip Clearance. Like any other cutting tool, there must be clearance behind the cutting edge before the drill can cut. This clearance can be readily seen on a properly-ground drill by using the drill-point gage, placing it over the heel of the point, as shown in Fig. 6. It will be noted that the angle here is 12 degrees less than the lip angle, and this is the proper clearance for the average drill

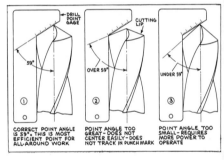

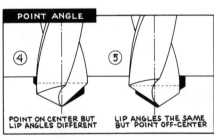

Figs. 1 to 5. Point angle and lip clearance are the two factors which enter into the grinding of drills. A point angle of 59 degrees and a lip clearance of from 12 to 15 degrees give best results for general work.

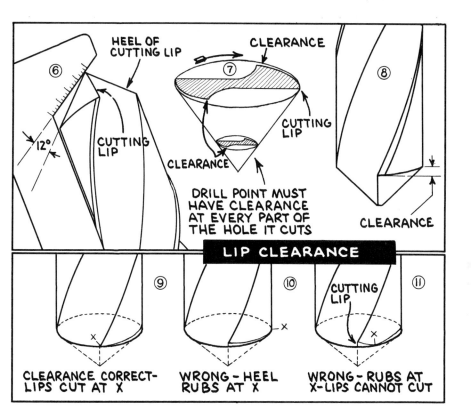

—from 12 to 15 degrees. Clearance can also be observed by holding the drill as in Fig. 8 and noting the difference between the lip and heel of the point. Two horizontal sections of a drill point are shown in Fig. 7. It can be seen that there must be clearance behind the cutting lips at every part of the conic recess which the drill cuts. With clearance properly ground on, the drill cuts at the cutting lips, leaving every part of the point behind the lips in the clear. Fig. 9 shows the correct clearance. Fig. 10 shows just the reverse of correct clearance—the drill rubs at the heel and the lips cannot cut. Likewise, if any part of the heel rubs against the conic recess, Fig. 11, the lips cannot cut.

Drill Grinding. With a fair understanding of point angle and lip clearance, the worker can now attempt to grind a drill. Even though the theory has been mastered, it is still somewhat of a trick to grind a drill offhand and arrive at the proper point. Experienced mechanics, through long practice, go through the motions almost mechanically, grinding points which are extremely accurate without the use of any mechanical guides or other aids. The worker who only occasionally grinds a drill should always use some form of guide. Figs. 12 to 16 inclusive

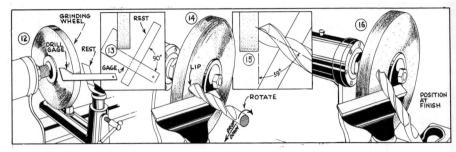

Figs. 12 to 16 show successive steps in sharpening drill on a wheel mounted in the lathe.

show one method of working. In this example, grinding is done on the lathe. the tool rest is first set by using the drill gage in the manner shown in Figs. 12 and 13. Now, if the lip of the drill is presented to the wheel while the body of the drill is at right angles to the rest, the point angle will be exactly as required, as can be seen in Figs. 14 and 15. From this starting position, the drill is rotated about one-sixth of a full turn, at the same time dropping the end about 12 degrees to give the required clearance. The proper swing is best acquired by swinging a properly-ground drill against the wheel, keeping the ground surface in contact with the wheel and noting the movements required to produce this surface.

Instead of dropping the drill to grind the lip clearance, the drill can be swung horizontally to produce the same effect. To grind drills in this manner, clamp a wood table to the tool rest of the grinder and on this table nail a guide block at an angle of 59 degrees with the side of the grind-

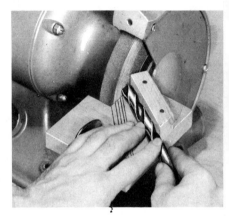

Fig. 17. A simple method of grinding utilizes a guide block to set the point angle.

ing wheel, as shown in Fig. 18. Mark off a series of parallel guide lines, each of these being on an angle of 12 degrees (the clearance angle) with the guide block, as shown. Now, if the cutting lip is placed against the side of the grinding wheel, with the body of the drill against the guide block, as shown in Fig. 19, the proper point angle will be obtained. From this position the drill is rotated about one-sixth of a full turn, at the same

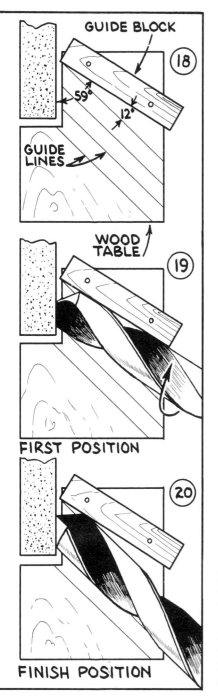

time moving to a position parallel with the penciled guide lines, as shown in Fig. 20. Each lip is treated in turn, checking with the drill gage to see that both are the same exact length. If desired, the grinding procedure can be reversed, starting at the position shown in Fig. 20. This has the advantage that the surface being ground can be seen at all times, but it has the disadvantage of producing a heavier burr at the cutting lips. In any case, care must be exercised not to rotate the drill too much, since over rotation will bring the lip on the opposite side into contact with the wheel, with the result that the grinding must be done all over again. One or two light twists on each lip will usually bring the drill to a sharp point. Touch-up grinding can be done by grinding the lips lightly and rotating about one-eighth of a turn for clearance. This will give a clean edge and can be done several times before the condition shown in Fig. 11 results, when the drill will demand grinding over the entire surface. Various mechanical drill grinders, both self-powered and for attachment to a standard grinder, can be purchased. Where considerable drill grinding must be done, it is usually a savings of both time and money to use one of these units. The usual device will handle drills from $\frac{1}{32}$ to ½-inch diameter.

Web Thinning. As shown in Fig. 23, the web of the drill becomes thicker as it approaches the shank. It follows, therefore, that the point of the drill becomes thicker as the drill is ground down and resharpened,

Fig. 21. Web thinning with round edge wheel.

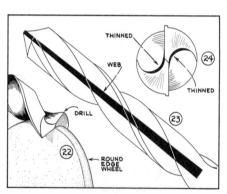

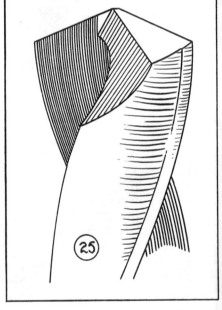

necessitating more power to force it through the work. To partly eliminate this heavy end thrust, the web of the drill should be thinned, as shown in Fig. 24. This operation is usually done on a round-face grinding wheel, the drill being held so that the wheel cuts in the flutes. The web can be thinned to about one-half its original thickness, providing the metal is removed from the immediate point only. Where the shop owner does not possess a suitable round-edge wheel, the web can be thinned on an ordi-

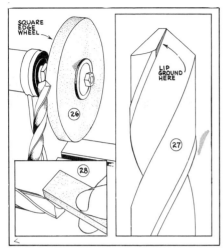

Figs. 26 to 28. Drilling in brass and other soft metal is best done with a drill ground in the manner shown.

nary square-face wheel. In this form of web thinning, most of the grinding is done on the back of the lips, the grind being carried up to the center of the point on each side, as shown in Fig. 25.

Drill for Brass. For drilling brass and copper it will be found advantageous to modify the cutting edges of the drill. The effect to be obtained is shown somewhat exaggerated in Fig. 27, which shows how the cutting edge of the lip is ground off. This makes the edge scrape rather than cut, and reduces the tendency of the drill to "dig in" in brass and other soft metals. This form of grinding can be done on a fine-grit wheel, as shown in Fig. 26. Very little metal is ground off, just a few thousandths of an inch, and for this reason it is often better to flatten the cutting edge with a small sharpening stone.

Special Grinding. While a point angle of 59 degrees and a clearance angle of from 12 to 15 degrees has been found best for average work, best efficiency is obtained if drills are specially-ground for the work to be done. For example, fiber takes a drill with a point angle of but 30 degrees, while manganese steel requires a 75-degree point angle. The countless drill-grinding variations now in general use are of undoubted value to the production shop worker, but are seldom useful in the home shop.

Wheels for Drill Grinding. As listed on page 94, the purified form of aluminum oxide (white in color) makes the best wheel for grinding high speed drills. It is not as tough as regular aluminum oxide, but runs cooler and is practically self-dressing. Second choice for high speed drills and first choice for general drill grinding is the regular type of aluminum oxide. In either case, the grit should be about No. 60 and the grade medium hard. The best wheel shape is a recessed center, but good work can be done on the side of a straight wheel.

CHAPTER TEN

Drill Grinding Attachment

The Drill Grinding Attachment (Fig. 1) is an accessory for sharpening drills of any size from ⅛ to ⅝ inch, using the standard tool grinder. It is a time saver in the hands of the expert mechanic, as well as a necessity for the inexperienced operator who could not otherwise expect to sharpen a drill accurately.

Simple adjustments are provided to produce the correct lip clearance and angles for quality or production drilling in any material.

The attachment is therefore extensively used by tool maintenance departments in production shops, as well as in small establishments where its low price and accuracy make it essential for keeping a lesser number of drills pointed.

Although the drill grinder attachment is designed for use on the Delta

Fig. 1

Fig. 2

7" tool grinder, by means of special adapters or holders it can be mounted to other grinders.

Wheel Dressing. Accurate drill grinding cannot be done unless the grinding wheel is true. Dressing the wheel before attempting to sharpen drills is of such importance that a Wheel Dresser Holder is included with the attachment. Fig. 2. The Wheel Dresser Holder will take dressing tools with shanks up to $5/16$ inches in diameter.

When dressing the wheel, the wheel dresser is rapidly passed across the face, moving the slide by hand. If moved too slowly the dresser will smooth the wheel and reduce its cutting qualities. The wheel is approached cautiously for the first cut, to remove the high spots and then fed only one or two knurl divisions of the screw for each pass across the wheel.

Mounting the Attachment. After the wheel has been dressed, the Wheel Dresser Holder is removed and a drill grinding attachment is mounted. By means of a set screw (Fig. 3) to take up any slack between the bracket and the slot and a hexagon head set screw the drill grinding attachment can be rigidly clamped to the grinder guard.

Adjustment for Lip Angle. Lip angles of 135, 118 and 82 degrees are most commonly used for drilling in various materials. These angles are

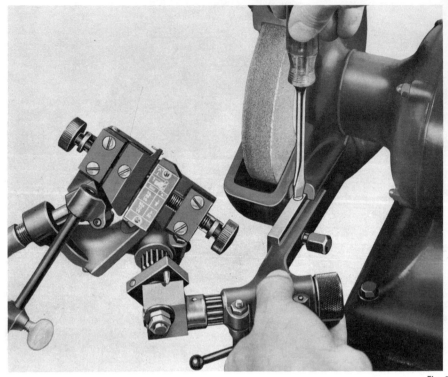

Fig. 3

marked on the scale which is mounted on the face of the feed screw bracket. Fig. 4. Intermediate angles may be set according to the operator's judgment.

Adjustment for Heel Angle. The drill grinding attachment is arranged to produce a 12 to 15 degree heel or clearance angle commonly recommended for 118 degree drills.

The heel angle can be decreased or increased by means of an adjusting screw.

The chart Fig. 5 illustrates some of the common lip and heel angles recommended for various materials.

Grinding a Drill. Ordinary carbon or high speed steel drills have 118 degree points which means that the angle between each cutting lip and the axis of the drill is 59 degrees. This shape is the most commonly used. The following are directions for clamping and sharpening such a drill. The same procedure should be followed for any other drill angle, after setting the angle scale to the required point.

There is an adjustment for the drill clamp jaws of the drill grinding attachment so that drills of all sizes can be held in the same line for uniform results.

Setting the Jaws. The right jaw of the drill grinding attachment remains fixed while the left jaw is moved by means of a knurled screw so that its left edge is set even with

Fig. 4

a graduation on the jaw scale, equal to the drill diameter.

The scale is divided into ⅛ inch intervals. For intermediate drill sizes, the proper setting between marks is estimated.

The drill is clamped in the jaws so that the drill flute is against a locating pin. The drill point projects ⅛ of an inch beyond the jaws. The required projection varies with the diameter of the drill. The exact setting can be determined from a table on the right jaw which can be seen in Fig. 3.

Adjustment for Grinding Position. The drill is brought to ¹⁄₁₆ inch of the face of the grinding wheel by cooperatively adjusting a series of controls and if necessary shifting the entire drill grinding attachment inward or outward.

The drill point is fed into the wheel by turning a feed screw thimble and by rocking the clamp jaws upward and downward by means of a long ball handle. One graduation of the thimble is fed for each pass up and down, using smooth, positive strokes. After enough material has been removed to clean up the lip and heel of the drill, the final thimble position is noted by reading its graduation with respect to a scribed line on top of the feed screw. The graduations have no meaning other than to control the

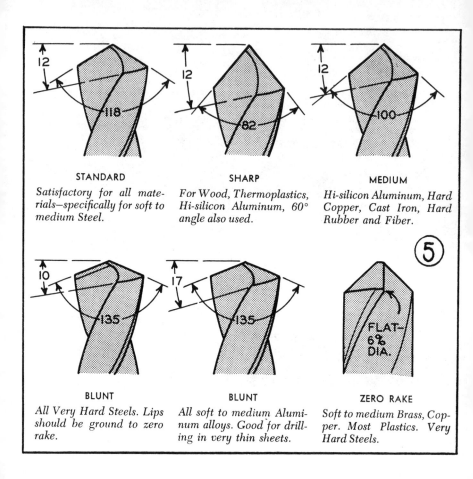

next grinding operation.

To grind the second lip equally with the first the drill must project the same amount from the jaws. A stop gage is used for this purpose. Fig. 6. The drill is rotated half a turn in the jaws positioning the second flute against the pin of the left jaw in the same manner as before, only this time keeping the shank of the drill in contact with the stop gage. This procedure brings a second lip in the exact position for duplicating the first lip.

Actually the entire operation of pointing a drill takes only a few minutes.

The operation of grinding three and four lip drills is the same as for two lip drills except that care must be taken so as to avoid grinding into the next flute when rocking the jaws.

Wheel Wear. When using the grinding attachment, grinding should not be limited to one side of the wheel face. The method described under adjustments for grinding position permits use of the entire width

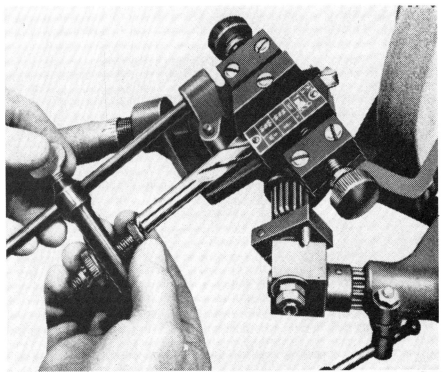

Fig. 6

of the wheel for uniform wear and longer wheel life.

The attachment will allow for wheel wear down to 5½″ diameter. As the wheel is dressed down, the mounting bracket moves farther into the supporting slot. A wheel too small for convenient drill grinding can be used for other work and should be replaced by a new full-sized wheel.

Web Thinning. After repeated sharpening more power is used in forcing the drill through the work. The web should therefore be thinned to reduce the required end thrust. The paragraph on web thinning on page 55 of this text should be followed to secure satisfactory drill performance.

CHAPTER ELEVEN

Buffing and Polishing

Polishing. Polishing is the general term applied to the complete process of removing tool marks, scratches, etc., from metals and other substances to produce a high-luster finish. The process is divided into three distinct parts. First of these is roughing. Roughing is done dry with abrasives in grit numbers from 40 to 80. Dry fining or fine wheeling, as the second operation is called, can also be done dry, but is often done on a greased wheel. Grits used are from No. 120 to 180. Finishing, also called oiling and buffing, is the final operation. It is done with fine grain abrasives combined with lard oil, tallow, beeswax, water, etc. The exact size of grain used in all operations will depend upon the original finish on the work and the desired finish on the completed product.

Polishing Wheels. The first operation of roughing can be done on a solid grinding wheel. However, because this is hard and has no flexi-

Fig. 1

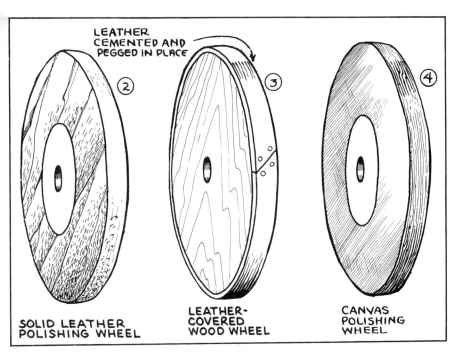

Figs. 2 to 4. Metals, plastics and lacquer-coated surfaces can be brought to a smooth and flawless finish by the application of suitably-graded abrasive grains carried on a polishing wheel.

bility, polishing is usually done on leather or canvas wheels made especially for this purpose. If the work is a flat surface, a solid leather or leather-covered wood wheel can be used; if the work is curved, it will demand a cemented canvas wheel or other type which has the required flexibility.

Setting Up Polishing Wheels. The old abrasive is first cleaned off by applying an abrasive stick about three numbers coarser than the abrasive on the wheel, as shown in Fig. 8. The wheel is then coated with glue, as in Fig. 5, after which it is rolled and pounded in the abrasive grains, as shown in Fig. 6. Ordinarily, one coat of abrasive is enough, but two or more coats can be applied to roughing wheels to lengthen their period of service. Each coat should be completely dry before the next is applied. After the wheel is dry, it should be balanced, as shown in Fig. 7. If any heavy spots are found, they should be corrected by nailing small pieces of lead to the wheel or by any other method which gives the desired result.

Concerning Glue. Animal hide ground glue is commonly used for applying abrasives to polishing wheels. It should be soaked in cold water from two to four hours, and is then brought to a heat of 140 degrees

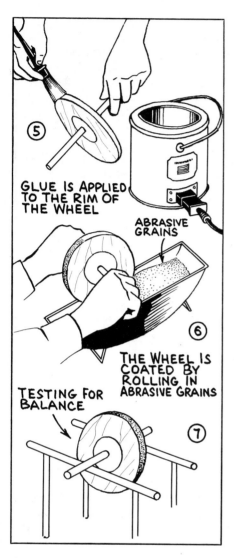

Fig. 8

F. in a suitable gluepot. Use definite weights of water and glue. Equal parts, by weight, is the right consistency for abrasive grains from No. 20 to No. 50. Finer grains demand a thinner glue. 60 to 70-grit takes a 40-60 mix (glue-water), 80 to 120-grit takes 33-67, while very fine abrasives from 150 to 220-grit will require 20 percent glue to 80 percent water. Wheels should set at least 48 hours before they are used. Instead of using glue, many workers prefer special polishing cements made for this purpose. These have the advantage of being already mixed, and require a shorter drying time.

How to Polish. The work is presented freehand to the wheel and on the outer quarter area of the wheel surface, as shown in Figs. 1 and 15. Avoid using too much pressure, as this tears out the abrasive grains and shortens the life of the wheel. Apply grease or other lubricant if required. An occasional application of lump pumice will clean the wheel if it becomes clogged. Work systematically over the area to be polished, inspecting the work frequently for defects which must be worked out.

Fine Wheeling. Fine wheeling is done the same as roughing except finer abrasive grains are used. Also, at this stage there is a greater use of the softer polishing wheels, and frequent use of grease or other lubricant to prevent the wheel from clogging.

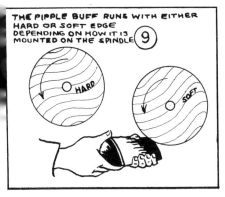

the lower side of the wheel, as shown in Fig. 18. Work presented as shown by the dotted lines, Fig. 18, will be torn from the hands with considerable force. The edges of the buff should be kept clean and round. Frayed edges can be dressed down with a buffing wheel rake, Fig. 16, while the buff is running. Any rough edge, such as a household food grater, can be used to dress buffs.

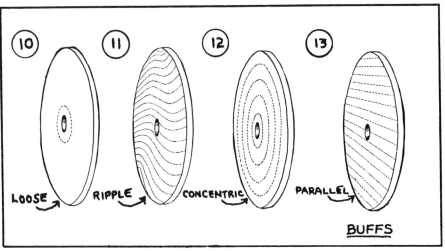

Figs. 10 to 13. *Buffing is the final polishing operation and is done with soft cloth or leather wheels.*

Buffing. Buffs are made of disks of muslin, felt, flannel, leather, etc., and are sewed in a wide variety of patterns as shown in Figs. 10 to 13, to produce hard or soft wheels. The loose buff (stitched once around the hole) as shown in Fig. 10, is a popular style. The ripple buff can be made to run with either a hard or soft edge by reversing it on the spindle, as shown in Fig. 9. As in polishing, the work must be done on

Strapping Belts. Belts of muslin, felt, leather, etc., can be run on the belt sander, as shown in Fig. 14, either with or without the backing plate depending on the nature of the work. In all respects, the belt can be treated the same as a polishing wheel, and can be coated with abrasive grains or buffing compound.

Buffing Compounds. Buffing compounds are various natural abrasives, such as emery, tripoli, pumice,

Fig. 14

crocus, lime and rouge, which are combined with a suitable wax or grease to form a mixture which can be readily applied to the revolving buff. The compound should be applied lightly and frequently to the buff as the work progresses, as shown in Fig. 19. The worker can make his own buffing compounds by melting beeswax in a double boiler, as shown in Fig. 20, and then adding the abrasive until a thick paste is formed. The molten mass is then poured into cardboard tubes, as in Fig. 21, or made into cakes, and when cold is ready for use. Very fine abrasives can be bonded with oil or water. The most common types of commercial compounds which are readily available are red rouge, pumice and tripoli.

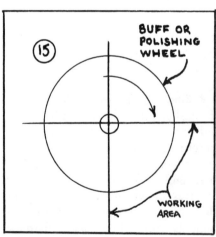

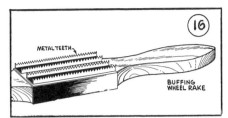

Fig. 19. The buffing compound is applied lightly and frequently to the revolving buffing wheel.

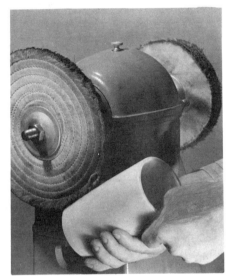

Fig. 17

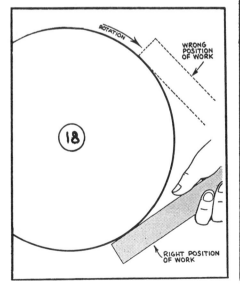

BUFFING AND POLISHING SCHEDULES

MATERIAL	METHOD OF WORKING
ALUMINUM	Polish at 5,500 s.f.m. using Nos. 80, 120 and 180-grits. All wheels over 120-grit should be well greased. Buff at 7,500 s.f.m., using tripoli for the first buffing and finishing with red rouge.
BRASS	Polish at 6,000 s.f.m. using Nos. 80, 120 and 180-grits. The 80-grit is necessary only for rough castings. Buff with tripoli or emery at a speed of about 5,500 s.f.m.
COPPER	Same schedule as brass. Fine-grit wheels should be greased. Avoid heavy pressure since copper heats quickly and holds heat longer than other metals.
CAST IRON	Use grits 120, 150 and 180. The two coarser grits can be run dry. Buff at 7,500 s.f.m. Buff with 220 to 240-grit silicon carbide applied to a greased rag wheel.
LACQUERED SURFACES	Use a lacquer suitable for buffing. Buff at 6,000 s.f.m., using any reliable brand of lacquer buffing compound.
NICKELED SURFACES	Buff at 7,500 s.f.m. using tripoli and lime. A perfect finish is necessary if the work is to be chromium plated.
PLASTIC	Polish with 280-grit silicon carbide. Buff with 400 and 500-grit silicon carbide on greased wheels. Finish with red or green rouge.
STEEL	Polish at 7,500 s.f.m., using aluminum oxide grits Nos. 90 and 120 dry and 180 greased. Buff with tripoli or a very fine grit aluminum oxide. For a mirror finish, buff with green rouge. For satin finish, buff with pumice on a Tampico brush.

Fig. 22

The red rouge buffing compound as the name implies is red in color and can be used on gold, silver and all precious metals to obtain a bright luster. Pumice is white in color and can be used on nickel, chromium, iron, stainless steel, cast brass and aluminum. Tripoli is brown and is ideal for articles made of pewter, brass, copper, aluminum, wood, plastic, horn and hard rubber. There is also a compound available which has an extremely sharp cutting action making it excellent for removing rust, scale and tarnish from tools and cast iron and steel. A buffing and polishing schedule is shown above.

Benzine or lacquer thinner will remove any film of compound left on the work after buffing.

CHAPTER TWELVE

How to Use Sanding Drums

Sanding Drums. Sanding drums of various sizes are extensively used for edge work, and can be satisfactorily worked on the drill press, lathe, flexible shaft, or direct-coupled to a motor shaft. The size most commonly used measures 3 inches in diameter and should be run at a speed of about 1800 rpm. Within reasonable limits, the higher the drum speed the smoother the finish. Excessive speed, however, causes overheating, and, where wood is being finished, the heat extracts a gummy pitch from the work which quickly clogs the abrasive sleeve.

Sanding on Lathe. Drums used on the lathe are fitted with taper shanks or screw-on fittings to permit fastening to the lathe headstock.

Sanding can be done freehand, but where edge work is being done, as in Fig. 1, a vertical support greatly simplifies the work. Drums with taper shanks should be safeguarded from

Fig. 1. A vertical sanding table clamped to the bed of the lathe permits accurate sanding of curved edges.

Figs. 2 & 3. Taper shank drums are securely held by means of a stud fitting through the headstock.

coming loose by supporting the end with the tailstock, using a sixty degree plain center. Another method of fastening uses a ¼ inch diameter stud which is turned into the hole in the end of the shank, the opposite end being held secure by means of a washer and wing nut, as shown in Figs. 2 and 3.

Narrow-Face Drums. Standard sanding drums measure 3 inches long and have a projecting nut on the free end. Narrow-face drums are 1 inch wide and are flush on the bottom, the tightening nut being on the shank end, as shown in Fig. 4. All standard operations can be done with these drums, the face width of 1 inch being sufficient to handle average ¾ inch thick work. The flush bottom also permits the finishing of inside corners, as shown in Fig. 7. When doing this kind of work, it is advisable to mount the sleeve so that it projects about $\frac{1}{32}$ inch beyond the bottom of the drum, as shown in Fig. 5, in order to prevent the drum bottom from burning the work. Narrow-face drums are fitted with a special tri-shape shank which permits mounting in either $\frac{5}{16}$ inch collets or three-jaw chucks, as shown in Fig. 6.

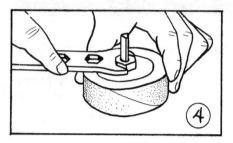

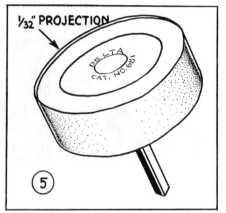

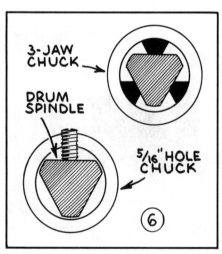

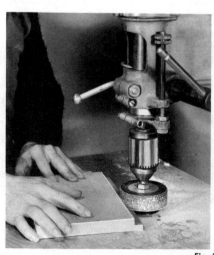

Fig. 7

Fig. 8. Using a sanding table.

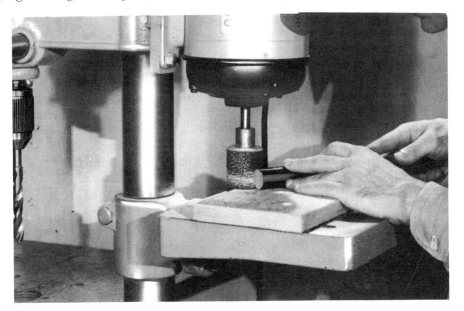

Fig. 9. Drum mounted direct to motor shaft.

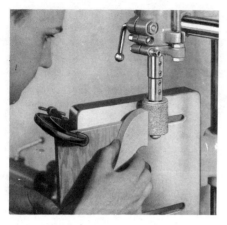

Fig. 10

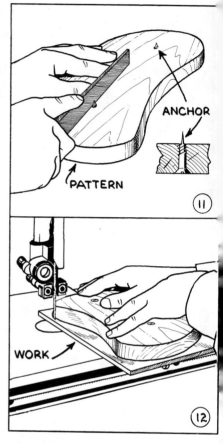

Sanding on Drill Press. Drums used on the drill press run in a vertical position most useful for average work. Good use can be made of fences, pivot pins, and other jigs to guide the work although most operations can be done freehand. Small drums will work inside the opening in the drill press table, while larger sizes can be worked by swinging the drill table to one side. Fig. 8 shows a useful sanding table. It has holes in it to accommodate the various size drums, so that the whole surface of the drum can be utilized by projecting the drum through the hole provided for it. Another worthy idea is shown in Fig. 9, which shows how a sanding drum can be carried on the lower end of the drill press motor spindle. A second table can be used, or the regular table can be swung around under the drum. Fig. 10 shows work being sanded with the drill table in a vertical position. Furniture legs which are to be fitted to a round column can be cut to shape by using a sanding drum in this manner.

Pattern Sanding. The most troublesome feature in edge sanding with a drum is that if the operator holds the work a moment too long at any one spot, the drum immediately cuts into the work, causing a ridge. This can be avoided and perfect work done if a pattern is used as a guide, as shown in Fig. 14. The pattern is a full-size template of the desired shape, with edges perfectly finished. It is fitted with two or more anchor points. These can be wood screws, with the projecting end filed to a thin, flat point, as shown in Fig.

Figs. 13 & 14. Pattern sanding eliminates ridge marks when irregular curves are being worked.

11. The work is fitted to the pattern, the anchor points holding it in place. Band sawing is then done, keeping about 1/16 inch outside the pattern, as shown in Fig. 12. Fig. 13 shows how the drill press table is fitted with a hardwood or metal ring. When the pattern is pressed against the collar, the drum cuts the work down to the same size as the pattern.

CHAPTER THIRTEEN

How to Use Cut-Off Wheels

General Use. Cut-off abrasive wheels in thicknesses from $\frac{1}{32}$ to $\frac{5}{32}$-inch are used for sawing all types of materials including metals, stone, and glass. The wheels are resinoid, shellac or rubber bonded to permit a slight measure of flexibility. The work can be done either wet or dry. The wet method is considerably faster and shows less wheel wear, but the dry method of cutting is quite suitable for average homeshop work. The surface speed of dry wheels should be about 7500 feet per minute, while wet wheels show best results at 6000 s.f.m. Extreme caution should be used in all operations involving abrasive cut-off wheels. Before attempting any cut-off operation, be sure that the abrasive wheel is properly mounted and fully guarded. Safety goggles should be used by the operator for added protection.

In every case, however, washers of blotting paper should be used on either side. These are generally furnished as part of the wheel. If extensive cut-off operations are to be performed use an abrasive machine designed to meet cut-off needs. Fig. 1 pictures an abrasive cut-off machine.

Cutting Thin-Wall Tubing. One of the most common uses of abrasive cut-off wheels is the cutting of tubing. In this operation, as done on the circular saw, it is advisable to hold the tubing in a suitable vee block, the block being held to the miter gage by means of the miter gage clamp attachment, as shown in Figs. 2 & 3. Notice that the work is supported on both sides, a slot in the block permitting passage of the wheel.

Cutting Solid Stock. Solid stock is cut in much the same manner as wood is sawed on the circular saw.

Fig. 1

Figs. 2 & 3. Cutting steel tubing with an abrasive wheel on the circular saw.

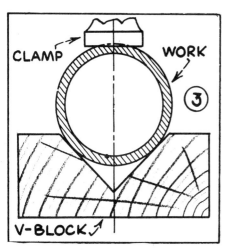

Some form of guide is always necessary. A suitable guard should be used. If a special abrasive wheel guard is not available, the regular saw guard can be fitted with a sheet metal hood to serve the purpose. Any metal cut with a dry wheel will discolor through heat generated by the wheel, but this surface film is easily removed by sanding with fine abrasive paper.

True Wheels Essential. A balanced wheel with a clean edge is necessary for good cutting. Wheels out of round will wobble and cut considerably wider than their own thickness. The use of a stick type dresser at regular intervals will keep the wheel in good condition. In using a stick dresser on cut-off wheels, do not be afraid to apply considerable pressure against the wheel. Wheel dressing is not a cutting operation, but a tearing away of the abrasive grains by pressing against the wheel with an abrasive which is somewhat harder than the wheel itself.

Cutting-Off on Grinder. Cut-off wheels can be used successfully on the grinder. The work must never, however, be worked freehand. A simple method of working is to turn the cen-

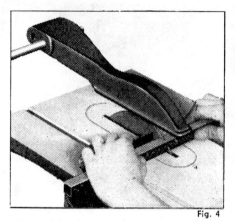

Fig. 4

Fig. 6

Fig. 5

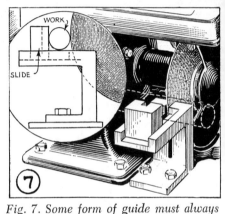

Fig. 7. Some form of guide must always be used when cutting-off on the grinder.

ter link of the tool rest upside down, Fig. 6, allowing it to pivot at the bottom. Work can then be clamped to the tool rest and swung into the wheel, as shown in Fig. 8. In another method of working, a table with a slot to accommodate a sliding block can be fitted to the tool rest or clamped to the workbench, as shown in Fig. 7. This provides a guide similar to the miter gage used on the circular saw. A simple method of wet cutting is shown in Fig. 8, and is quite clean in operation. The water pan is a bread pan, obtainable at any dime store. The guard plate is used in actual operation and is removed in the picture only to show how the bottom of the wheel runs in the coolant. The use of the grinder for cutting-off has one serious drawback in that the projection of the motor limits the size of stock which can be worked to about ½ inches square. The belt-drive grinder, however, has much more clearance.

Diamond Blades. Glass, hard alloys and gems are commonly cut with the use of a diamond cut-off wheel. This is a metal wheel with diamond chips impregnated around the rim.

Fig. 8. A bread pan fitted below the grinder holds the coolant when wet cutting or grinding is to be done.

Ready-made wheels, six inches in diameter, cost about $5. One carat of diamond bort (value about $2) will charge the home-made wheel. This is made from a disk of 1/16 inch thick steel. Nicks cut around the wheel receive the bort mixed with vaseline, the chips then being sealed in place by rolling with a hardened steel roller, as shown in Figs 9 & 10. Diamond blades are generally run at somewhat lower speeds than other types of cut-off wheels, and must always be run wet. The grain size for average work should be 50-grit.

Miscellaneous Cut-Off Wheels. The general rule is to use the hardest of wheels for soft materials and softest of wheels for hard materials. The hardness or grade of wheel should be such that the bond on the wheel breaks down at the same speed at which the abrasive grit in the wheel becomes dull thus avoiding wheel "loading" or glazing. There are so many elements involved in selecting blades such as the material, shape, speed, etc., that a reliable abrasive product concern should be consulted before selecting abrasive cut-off wheels.

Cutting-Off on Lathe. The lathe performs excellently as a cut-off machine, especially on small work which permits the use of the slide rest for feeding, as shown in Fig. 11. Either the slide rest feed can be used or the

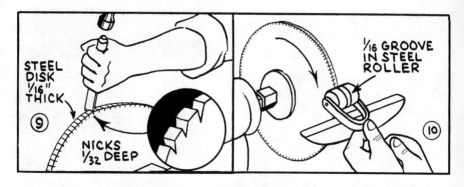

Figs. 9 & 10. Making a diamond blade.

Fig. 11. Cutting-off on the lathe.

work table can be advanced by hand, guided by the bar which works in the slot of the slide rest. A special table can be made if desired so that work can be cut at the top of the blade, much the same as on the circular saw. Wet cutting is more readily done on the lathe than on any other machine.

CHAPTER FOURTEEN

Miscellaneous Abrading Operations

Tumbling. Tumbling is extensively used in industry for finishing small metal parts. The process can be worked successfully in the small shop, using a wooden barrel driven on the outboard end of the lathe. The barrel should be made from hard maple, as shown in Fig. 2, and must be solidly constructed, especially if castings of fair size are to be tumbled. If tumbling is done wet, as is sometimes the case, the barrel must also be watertight. One section is hinged to form a lid. The barrel is fastened to a 15-inch diameter hardwood pulley.

The work is generally processed through a cycle of at least three operations. The first is an ashing or scouring operation which removes all tool marks and roughness on the surface of the article and renders it smooth. The second stage is the burnishing or semi-polishing operation, while the final stage is the true polishing operation. The time required to tumble finish runs from 16 to 48 hours for the three operations. The first stage of the process demands the most time; the final polish can be done in a very short time providing the previous abrading has been done thoroughly. The abrasive used is in loose grain form, the grit being somewhat coarser than used in wheel polishing. Nos. 24 to 36 are used for roughing; 40 to 80 for semi-polishing; 100 and 120 for polishing. The exact abrasive varies considerably, depending upon the work being tumbled. The barrel is loaded between one-third and two-thirds full, the work comprising about 60 percent of the load while the balance is made up with the abrasive grains and small pegs of the material being tumbled. Slow speed is absolutely essential and should never exceed 90 rpm. A suitable drive which does not interfere with the lathe drive proper is shown in Fig. 3.

Spun Finish. An attractive finish

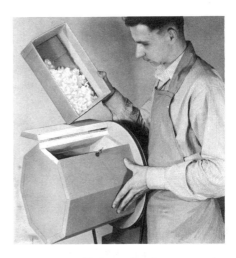

Fig. 1. Tumbling with abrasive grains makes a good finish for metal or plastic parts. A cycle of three operations is generally required.

81

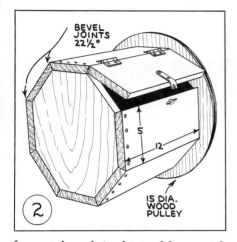
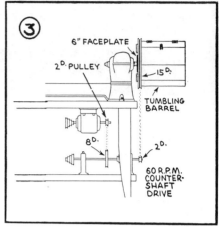

for metal work is obtained by scratch brushing at slow speed, as shown in Fig. 4. A stick coated with coarse abrasive grains can also be used. The finish consists of a series of minute rings spun around the work, the depth of the serrations depending upon the softness of the metal itself and the grit of the abrasive used. This abrasive treatment is extensively used as a finish for spun aluminum projects.

Grinding Glass. Glass edges can

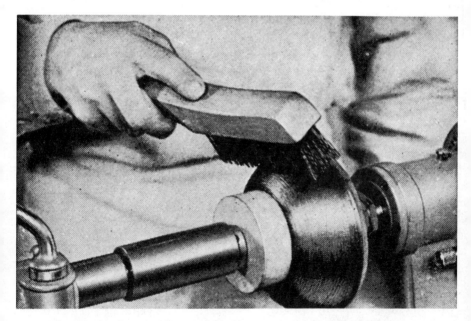

Fig. 4. Scratch brushing at slow speed produces a serrated finish. A high speed gives a satin finish.

Fig. 5. Grinding glass edges.

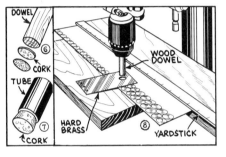

Figs. 6 to 8 show various methods used in producing an engine finish.

be ground to an almost perfect polish with the use of silicon carbide abrasive papers used on the disk sander, as shown in Fig. 5. Standard 9 by 11 inch sheets of wet-or-dry paper can be cut into disks for occasional work of this nature. 120-grit gives a smooth, mat edge, while 220 to 320-grit brings up a very good polish. The glass should air cool between cuts.

Engine Finish. An engine or spot finish is produced on the drill press by the methods shown in Figs. 6, 7 & 8. Fig. 6 shows how a dowel stick is capped with an abrasive disk for this purpose. Instead of using an abrasive disk, the tip can be leather and the abrasive bonded with wax or grease and fed to the work. The drill press should run about 1200 rpm. Solid abrasive sticks can be used, but require careful alignment. A tiny cup wheel, Fig. 7, makes a ring pattern. Fig. 8 shows a good method of working the engine finish. A softwood dowel is used. A strip of thin, hard brass with a hole through it is located on the drill table immediately below the dowel. The work is fed with a mixture of abrasive grains and oil or water, the dowel abrading this into the metal.

Internal Grinding. Small holes can be ground by mounting an abrasive stick in a chuck held in the lathe tailstock, as shown in Fig. 9. The set-over tailstock sets the depth of cut. A small hand grinder can be used for the same job, and also for a wide variety of other work.

Other Lathe Operations. Other examples of grinding jobs which can be done on the lathe are shown in Figs. 11 & 12. Fig. 11 shows a groove being run in on a bushing, using a round edge wheel. The same set-up with straight wheel and slide rest feed is used for surfacing cylinders. Fig. 12 shows a flat being worked on the end of a shaft. The lathe does not turn in this instance, but is locked by means of the index pin.

Drilling Glass. The drilling of glass is an operation that may be done with ease on the drill press, although difficult by any other method. The drill employed is a piece of brass tubing with an outside diameter equal to the

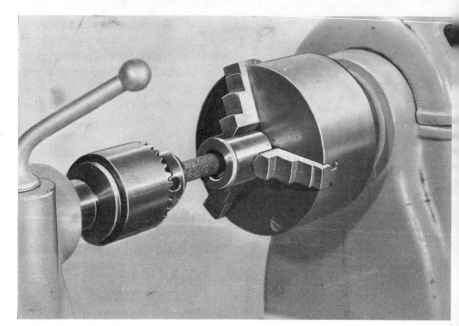

Fig. 10. Tool post grinders worked on the slide rest are used for a wide variety of lathe grinding jobs.

Fig. 11

Fig. 12

size of the hole to be drilled. The tubing should be slotted with one cut, using a very narrow saw. The cut need not extend more than about ¼ inch from the end of the tube. Similar results are effected by notching the end of the tube in two or three places. The tube is not sharpened in any way —it is simply cut square on the end, and then slotted or notched as men-

Fig. 13. Drilling glass on the drill press.

tioned. The glass should be supported on a perfectly flat piece of wood, or, better, on a piece of felt or rubber. A dam of putty is built around the place where the hole is to be drilled, or a felt ring can be used for the same purpose. The well is fed with a mixture of 80-grit silicon carbide abrasive grains combined with machine oil or turpentine. A carbide tip drill could be used to do this operation without the use of any abrasive mixture, but because of occasional use and cost, the craftsman would be better off using the brass tube abrasive method.

Grinding Keyways. Keyways and similar work can be done with small abrasive wheels mounted in the drill press. The best method of holding and feeding the work is to employ the lathe slide rest. This is easily fitted to

Fig. 14. The carbide tipped drill and the slotted brass tube are excellent for drilling holes in glass.

Fig. 15

Fig. 16

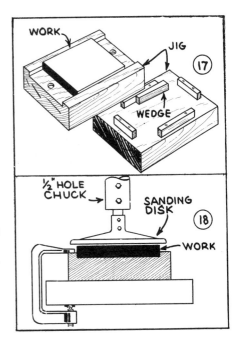

Figs. 15 & 16 show use of slide rest in drill press grinding.

the lathe table by means of the same bolts used to attach it to the lathe. An under view, showing the slide rest being fitted, is shown in Fig. 15. Fig. 16 shows the keyway being cut. A speed of 5000 rpm. should be used. Wheel shapes in any size, abrasive, or grit can be obtained ready mounted on shanks which can be held in the drill press chuck.

Surface Grinding. A 3-inch cup wheel worked in the drill press offers one of the best methods of surface grinding. The work can be held by the lathe slide rest mounted as already described or in any suitable jig. The slide

Fig. 19. A cup wheel mounted in the drill press will handle a wide variety of surfacing operations.

Fig. 20

rest permits feeding the work to the wheel, while work held stationary in a machinist's vise or simple wood fixture, Fig. 17, is surfaced by feeding the wheel to the work. Fairly large surfaces can be covered by using a drill press column collar and swinging the work below the abrasive wheel, the same method as described for grinding jointer knives (see page 37). Large work can also be handled by using a sanding disk in the manner shown in Fig. 18.

Grinding on Shaper. Sanding and grinding can be done on the wood shaper, using sanding drums, straight wheels, cup wheels, and any other form of grinding stone which can be fitted to the spindle. For most work, the speed is excessive and should be reduced to about 5000 rpm. This can be done by belting the shaper pul-

Fig. 21

Figs. 20 and 21 show the shaper being used for grinding operations. Almost any tool can be adapted for abrasive work.

ley to a pulley fitted to the lower end of the drill press motor, lowering the drill head to suit. Figs. 20 & 21 show typical set-ups. The operation shown in Fig. 21 requires the use of a wood table to bring the work approximately

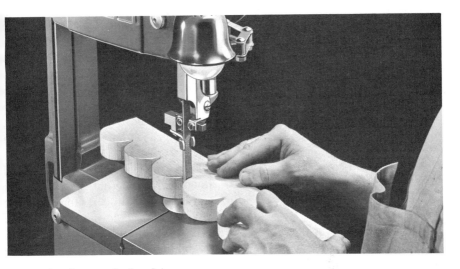

Fig. 22. Sanding on the band saw.

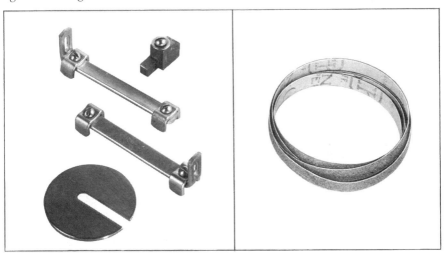

Fig. 23. The band saw sanding attachment and abrasive belt.

level with the top of the cup wheel which is being used. Final adjustment as well as feed is provided by the standard travel of the shaper spindle.

Sanding on Band Saw. The use of narrow abrasive belts on the band saw provides an excellent method of sanding edge work. Belts up to 1 inch wide can be used. Guides are provided for some band saws to permit the use of belts, but on other saws the worker can improvise simple guides. As far as guides are concerned, the belt will work perfectly well without them. Ribbon abrasive belts in all standard widths and grain sizes are obtainable and are easily spliced to suit the size of the band saw. Perfect tracking is assured by the upper tilting wheel of the saw, while the standard band saw

Fig. 24

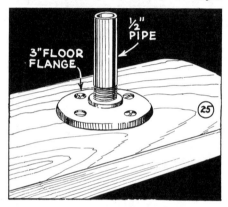

adjustment permits tensioning.

Sanding on Lathe. All operations described in the chapter on the disk sander can be worked with equal facility on your lathe with the addition of a sanding attachment, Fig. 24. The attachment consists of a sanding table, miter gage, sanding disk and abrasive disks.

The commercial sanding table is supported, both front and back, and offers perfect support for work. It raises and lowers to suit the operation and is graduated to show angle of tilt. The table has a milled slot for a miter gage which assures you of accurate sanding operations.

For the craftsman who desires to make his own, a wood table as shown in Fig. 25 will serve the purpose. The wood top is fitted with a pipe shank which can be accommodated in the tool rest base. A groove to take the miter gage can be run in on the table

Fig. 26 shows the homemade sanding table in use.

surface if desired. One disadvantage of the home-made type is that it will not tilt for sanding compound miters.

Sanding on Other Tools. Almost any shop tool can be used for sanding and grinding. Sanding disks work perfectly when mounted on the circular saw arbor; small sanding drums are commonly used on the scroll saw; simple jigs can be made for the lathe to permit the use of sanding belts. Where the work cannot be taken to the machine, a flexible shaft suitable for handling sanding disks and grinding wheels can be used to advantage for many operations.

CHAPTER FIFTEEN

Excerpts from Your

Delta Craftsheet

The Informative material on the following pages has been reprinted from Delta Craftsheets devoted to abrasive tool operations.

Delta Craftsheets appear in each issue of the bi-monthly publication, the Deltagram, and complete sets of craft sheets are available for your woodworking library. Each sheet is devoted to a series of hints concerning power tool operations and general woodworking know-how that every home workshop enthusiast will find valuable.

SPARK TEST FOR METALS

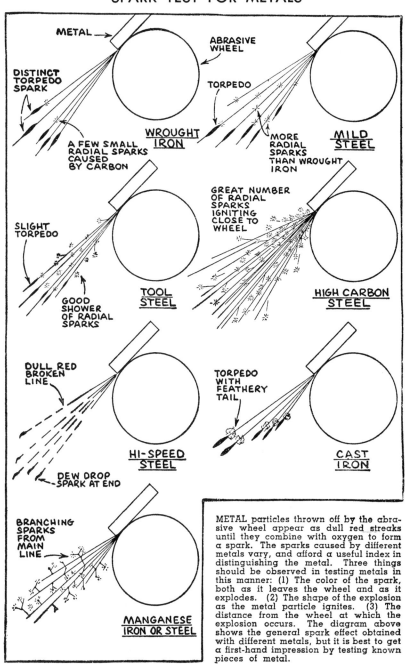

METAL particles thrown off by the abrasive wheel appear as dull red streaks until they combine with oxygen to form a spark. The sparks caused by different metals vary, and afford a useful index in distinguishing the metal. Three things should be observed in testing metals in this manner: (1) The color of the spark, both as it leaves the wheel and as it explodes. (2) The shape of the explosion as the metal particle ignites. (3) The distance from the wheel at which the explosion occurs. The diagram above shows the general spark effect obtained with different metals, but it is best to get a first-hand impression by testing known pieces of metal.

ABRASIVES AND ABRASIVE TERMS

Aluminum Oxide... An artificial abrasive, with a base of the natural clay-like mineral, Bauxite. Color, brown-gray to black; in pure form, white. Hardness, 9.*

Bond.. Clay or other substance which holds abrasive grains together to form wheels, etc.

Crocus.. A natural abrasive formed from oxide of iron. Color, purple. Used for fine polishing. Hardness, 2.

Emery... A natural abrasive, being an impure form of crystalline alumina. Much softer than aluminum oxide. Color, dull black. Hardness, 8.

Garnet. A natural abrasive mined in the United States. Extensively used in woodworking. Color, red. Hardness, 7.

Glazed.. Said of a wheel or stone which has become clogged with metal particles so that it will not cut.

Grade.. The resistance of the bond in a grinding wheel to any force tending to pry the abrasive grains loose. Has nothing to do with the hardness of the abrasive itself.

Grit.. The size of the abrasive grains, determined by the number of grains which, end to end, equals one inch.

Lime.. A fine natural abrasive, used extensively in the final polishing of brass and nickel. Color, white. Hardness, 1.

Oilstones... General descriptive term applied to all abrasives when made into stones for bench use.

Pumice... A natural abrasive Used for final polishing, cutting down finishing coats of varnish, etc. Several grades of fineness. Color, off-white. Hardness, 4.

Quartz... Commonly called Flint. A natural abrasive. Least expensive of all abrasives but very soft. Color, yellow. Hardness, 6.

Rottenstone... A natural abrasive. Negligible cutting action but good polisher. Color, off-white. Hardness, 3.

Rouge... A natural abrasive in powder form. Graded fine, very fine, extra fine. Color, red; also, green (chromium oxide). Hardness, O.

Silicon Carbide... An artificial abrasive made by fusing silica sand. Color, gray, green or black. Hardness, 10.

Structure.. The spacing of abrasive grains in a grinding wheel. Usually represented by a number from 1 to 12, the smaller numbers indicating close spacing of grains.

Tripoli.. A silicious powder consisting of tiny skeletons. Color, pink. Graded. Used in fine polishing. Hardness, 5.

Numbers given in this table list the hardness of the abrasives in sequence, from 10 (the hardest) to 0 (the softest). The number has no grading value, but is simply used to indicate the hardness of the abrasives in descending order.

COATED ABRASIVE SELECTION

MATERIAL	ABRASIVE	ROUGH	FINISH	FINE
Hard Woods	Garnet or Alum. Oxide	2½-1½	½-1/0	2/0-3/0
Soft Woods	Garnet	1½-1	1/0	2/0
Aluminum	Alum. Oxide	40	60-80*	100
Bakelite	Alum. Oxide	36-40	60-80	100
Cast Brass	Silicon Carbide	36-40	60-80	80-120
Comp. Board	Garnet	1½-1	½	1/0
Copper	Alum. Oxide	40-50	80-100	100-120
Cork	A. O. or Garnet	3	1	1-0
Fiber	Alum. Oxide	36	60-80	100
Glass	Silicon Carbide	50-60	100-120	100-320
Horn	Garnet	1½	½-1/0	2/0-3/0
Iron (Cast)	Silicon Carbide	24-30	60-80	100
Ivory	Alum. Oxide	60-80	100-120	120-280
Paint (removing)	Flint	3-1½	½-1/0	...
Plastic	A. O. or Garnet	50-60	120-180	240
Steel	Alum. Oxide	24-30	60-80	100

COMPARATIVE GRAIN SIZES

NO.	GARNET	FLINT	EMERY	NO.	GARNET	FLINT	EMERY	NO.	GARNET	FLINT	EMERY
400	10/0	...	...	150	4/0	2/0	½	40	1½	2½	...
320	9/0	...	FF	120	3/0	1/0	1	36	2	3	...
280	8/0	...	F	100	2/0	½	1½	30	2½	...	...
240	7/0	...	3/0	80	1/0	1	2	24	3	...	...
220	6/0	4/0	2/0	60	½	1½	2½	20	3½	...	...
180	5/0	3/0	1/0	50	1	2	3	16	4	...	...

GRINDING WHEEL SELECTION*

WORK	ABRASIVE	GRIT	GRADE	BOND
Aluminum (surfacing)	Alum. Oxide (White)	46	Soft	Vitrified
Aluminum (cutting-off)	Alum. Oxide	24	Hard	Resinoid
Brass (surfacing)	Silicon Carbide	36	Medium	Vitrified
Brass (cutting-off)	Alum. Oxide	30	Very Hard	Resinoid
Cast Iron	Silicon Carbide	46	Soft	Vitrified
Chisels (woodworking)	Alum. Oxide	60	Medium	Vitrified
Copper (surfacing)	Silicon Carbide	60	Medium	Vitrified
Copper (cutting-off)	Silicon Carbide	36	Hard	Rubber
Cork	Alum. Oxide (White)	60	Soft	Vitrified
Cutters (moulding)	Alum. Oxide	60	Medium	Vitrified
Drills (sharpening)	Alum. Oxide (White)	60	Medium	Vitrified
Glass (grinding)	Silicon Carbide (Green)	150	Hard	Vitrified
Glass (cutting-off)	Silicon Carbide (Green)	90	Hard	Rubber
Glass (cutting-off)	Diamond	60	Medium	Copper
Leather	Silicon Carbide	46	Soft	Vitrified
Plastic	Silicon Carbide	60	Medium	Rubber
Rubber (hard)	Silicon Carbide	46	Medium	Resinoid
Saws (gumming)	Alum. Oxide	60	Medium	Vitrified
Steel (soft)	Alum. Oxide	60	Medium	Vitrified
Steel (high speed)	Alum. Oxide (White)	60	Soft	Vitrified
Tile (cutting-off)	Silicon Carbide	30	Hard	Resinoid
Tubes (steel)	Alum. Oxide	60	Hard	Rubber
Welds (smoothing)	Alum. Oxide	36	Hard	Vitrified
Wood (hard)	Silicon Carbide	30	Soft	Vitrified

* Adapted from tables by The Norton Company.

Recommended WHEEL SPEEDS

Chisel Grinding	5,000-6,000 s.f.m.
Cut-off Wheels	6,000-8,000 s.f.m.
Surface Grinding	4,000-6,000 s.f.m.
Polishing	6,000-9,000 s.f.m.
Polishing (soft rubber wheels)	4,000 s.f.m.
Buffing	6,000-9,000 s.f.m.
Scratch Brushing (rough finish)	600 r.p.m.
Scratch Brushing (satin finish)	4,000-6,000 s.f.m.
General Grinding	5,000-6,500 s.f.m.
Internal Grinding	2,000-6,000 s.f.m.

Recommended BELT and DRUM SPEEDS

48 inch abrasive belts	3,100 s.f.m.
6 to 10-ft. abrasive belts	2,800 s.f.m.
10 to 16-ft. abrasive belts	2,400 s.f.m.
48 inch polishing belts	4,000 s.f.m.
3 inch drums (coarse grit abrasive)	1,800 r.p.m.
3 inch drums (fine grit abrasive)	2,400 r.p.m.
1 inch drums (closed coating)	1,200 r.p.m.
1 inch drums (open coating)	1,800 r.p.m.
10 to 12 inch abrasive disks	1,800 r.p.m.
Abrasive disks	4,500 s.f.m.

GRINDING WHEEL SPEEDS IN R. P. M.

DIAMETER OF WHEEL	R. P. M. FOR STATED SURFACE SPEED							
	4000 s.f.m.	4500 s.f.m.	5000 s.f.m.	5500 s.f.m.	6000 s.f.m.	6500 s.f.m.	7000 s.f.m.	7500 s.f.m.
1	15,279	17,189	19,098	21,008	22,918	24,828	26,737	28,647
2	7,639	8,594	9,549	10,504	11,459	12,414	13,368	14,328
3	5,093	5,729	6,366	7,003	7,639	8,276	8,913	9,549
4	3,820	4,297	4,775	5,252	5,729	6,207	6,685	7,162
5	3,056	3,438	3,820	4,202	4,584	4,966	5,348	5,730
6	2,546	2,865	3,183	3,501	3,820	4,138	4,456	4,775
7	2,183	2,455	2,728	3,001	3,274	3,547	3,820	4,092
8	1,910	2,148	2,387	2,626	2,865	3,103	3,342	3,580
10	1,528	1,719	1,910	2,101	2,292	2,483	2,674	2,865

HOW TO SHARPEN CHISELS

Wood Chisels. Wood chisels should be hollow ground. Project the chisel straight into the wheel to remove nicks, as shown in Fig. 1; then, adjust the tool rest or the chisel grinding attachment to the required position to grind the bevel, Figs. 2 and 3, working the chisel squarely across the face of the wheel, as shown in Fig. 4. Worked on the face of the wheel, the

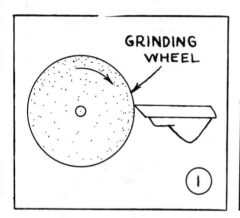

Fig. 1. Remove nicks by pushing chisel straight into wheel.

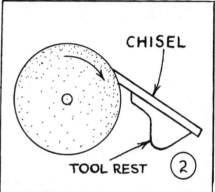

Fig. 2. Tilt tool rest to grind bevel to the required angle.

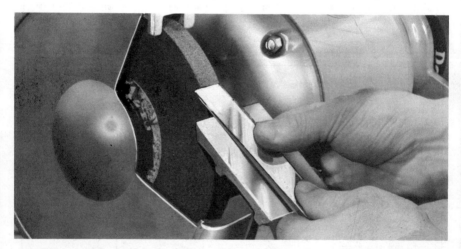

Fig. 3. Tool grinding is done on an aluminum oxide wheel, with the edge of the tool against the direction of rotation.

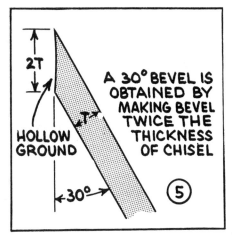

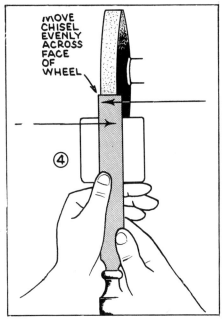

bevel will have a slight hollow, making it easy to hone to a perfect edge several times before regrinding again becomes necessary. The bevel should be about 30 degrees, this being obtained by making the bevel twice the thickness of the chisel, as shown in Fig. 5. A 20 degree bevel can be used for softwood, but this thin wedge will crumble on hardwood, as pictured in Fig. 6.

Honing. Either an aluminum oxide or a silicon carbide oilstone will give good results in honing or whetting the chisel edge after grinding. The sharpening stone should always be oiled, the purpose of this being to float the particles of metal so that they will not become embedded in the stone. Use a thin oil or kerosene. Wipe the stone after using. Honing is necessary because grinding forms a burr at the chisel edge, as shown in Fig. 7. To remove the burr, place the chisel diagonally across the stone as shown in Fig. 8, and stroke backward and forward, bearing down with both hands. The heel of the chisel should be a slight distance above the surface of the stone, as shown in Fig. 9. Next, turn the chisel over and stroke the back on the stone, making certain to keep the tool perfectly level, as shown in Fig. 10. Alternate the honing on bevel and back until the burr is completely removed.

It will now be noted that honing puts a secondary bevel on the chisel, and this is the correct technique for chisels, plane irons, knives, etc. This method gives a clean edge with a minimum amount of labor. When the honed bevel becomes too long through repeated whettings, Fig. 12, chisel should be reground. Figs. 13 and 14 picture the common method of hand honing. Fig. 15 shows standard test for sharpness—the chisel should "bite" on the thumb nail.

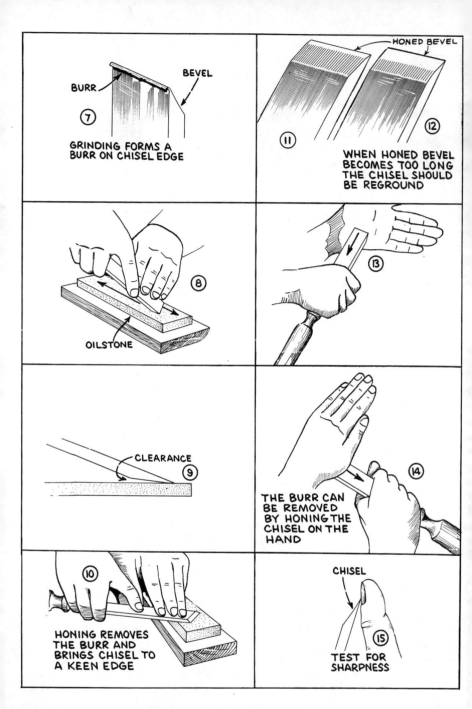

Figs. 7 to 15. Drawing shows honing methods.

CHAPTER SIXTEEN

Helpful Hints About Machines and Accessories

What to Look for When You Buy Abrasive Tools

Abrasive tools are important machines in the workshop. They eliminate much of tedious hand finishing otherwise necessary to give your projects that professional touch. In addition, they keep your cutting tools sharp for top quality performance. Listed below are a number of characteristics and operating features which apply to every model and should be considered when buying any basic abrasive machine.

Established Manufacturer. When buying abrasive tools it is wise to choose those made by a manufacturer who has established through the years a record of quality and reliability in the production of precision engineered power tools for the industrial and consumer markets. Thus you will be sure of getting top-quality tools that will last a lifetime. It costs very little more at the start and much less in the long run to equip your workshop with the best in power tools.

Availability of Replacement Parts. When you choose the product of an established manufacturer you can be sure that replacement parts will always be available no matter how old your tools may be.

Availability of Accessories. Check to make sure that the manufacturer of the abrasive tools also produces a complete line of accessories such as grinding wheels, abrasive disks and belts, various attachments for drill grinding, etc. Thus you will be certain that the accessories you buy will fit your machines with no necessity for makeshift arrangements.

Solid Construction. All abrasive tools should be solidly constructed to provide the strength and rigidity needed for long life and continued precision accuracy.

Sturdy Tilting Tables. On belt and disk sanders the tables should be precision ground to very close tolerances in addition to being heavily ribbed to give ample support. A miter gage slot should be accurately machined parallel with the sanding surface so that a miter gage can be used with jigs and fixtures with unerring accuracy in sanding operations. The tables should also be of the tilt type, tilting at least 45° downward and 20° upward.

Adjustable Tool Rests. On all grinders tool rests should be fully adjustable horizontally as well as vertically so that full advantage can be taken of both sides of the grinding wheel to assure uniform wheel wear.

Safety Features. The on-off switch on grinders should be located on the front of the machine conveniently placed to avoid any unnecessary movement to quickly stop the grinder should it be required. Provisions should be made for mounting wheel guards so they can easily be mounted should they not be included with the purchase price of the machine.

Correct Speeds. Grinder speeds should range from 5000 to 6500 feet per minute to meet recommendations set forth by manufacturers of abrasive products. Although speeds vary for different jobs, most operations can be performed satisfactorily within this range.

Proper Accessories Enable You to Do More Jobs . . . And to Do Them Easier and Faster

THROUGHOUT THE PRECEDING PAGES, AS PART OF THE DESCRIPTIONS OF HOW TO PERFORM DIFFERENT OPERATIONS ON YOUR ABRASIVE MACHINES, VARIOUS ACCESSORIES WERE MENTIONED AND PICTURED.

FOR YOUR CONVENIENCE, EACH OF THE IMPORTANT ACCESSORIES AVAILABLE FOR THE BASIC ABRASIVE MACHINES IS LISTED AND DESCRIBED BELOW. MORE COMPLETE SPECIFICATIONS AND CATALOG NUMBERS MAY BE FOUND IN THE LATEST DELTA CATALOG AS AVAILABLE FROM YOUR NEAREST DELTA DEALER, OR DIRECTLY FROM THE DELTA POWER TOOL DIVISION, ROCKWELL MANUFACTURING COMPANY, 400 NO. LEXINGTON AVENUE, PITTSBURGH 8, PENNA.

Drill Grinding Attachment. Anyone can do accurate drill sharpening using the drill grinding attachment. No special skill is required. It is designed for use on the Delta 7" Tool Grinder and will sharpen drills ranging in size from 1/8" to 5/8", assuring proper lip clearance and angles to any degree needed. The drill grinding attachment is excellent for use by tool maintenance departments, production shops, etc., where it is essential to keep all drills sharp continually.

Plane Blade Grinding Attachment. The plane blade grinding attachment provides the secure bearing needed to sharpen plane blades, hand chisels, scrapers, etc. on your grinder. Fully adjustable, it should take knives up to 3 3/16" wide, permitting them to be ground to any desired angle.

insure the safety of the operator. The deluxe type grinding shield illustrated in addition to furnishing full protection to the operator has a built-in lighting system which results in more efficient and safer operation of the grinder.

Abrasive Disks and Disk Adhesive for Disk Sanders. Aluminum oxide and garnet sanding disks are available in various grits and grades enabling you to handle all type materials on your present machine. To avoid any delay in change over or disk renewal a special quick acting disk adhesive should be used. No messy gluing should be required. With the proper adhesive, disks can be removed and replaced every few minutes if necessary.

Diamond Pointed Wheel Dresser and Tool Holder. In order to do accurate grinding, grinding wheel faces must be level and true. The diamond wheel dresser will dress all types of general grinding wheels. To facilitate ease of operation of this tool, a special wheel dresser tool holder is available to provide the secure bearing needed in dressing operations.

Grinding Wheels. Aluminum oxide grinding wheels of various grits and grades are available for all types of grinding operations. They should be fully balanced to run absolutely true and vibrationless.

Buffing, Wire and Fibre Wheels. Wire and fibre wheels are used to remove rust and scale from metals in general cleaning operations while buffing wheels are used in polishing operations.

Abrasive Belts for Belt Sanders. Garnet belts, for woods, and aluminum oxide belts for metals are available for your belt sander in various grits and grades. They should be cloth backed to withstand all types of abuse in miscellaneous sanding operations.

Grinding Shields. Safety shields should be available for all grinders. They are important accessories that

Dust Collectors for Abrasive Tools. Both the operator and valuable machinery should be fully protected from harmful metal and dust particles produced in the various abrading operations. Dust collectors are ideally suited to solve this problem and are available for most types of abrasive machines. All units should be self contained and come completed with attachment packages specifically designed to fit the individual machines.

Miter Gage for Belt and Disk Sanders. The miter gage should be used whenever possible in sanding plain and compound miters to assure uniformity and accuracy. It should be

possible to make exact settings for angles ranging from 0° to 60°, both right or left.